I0471446

Viscoelastic Duende of Life

Evolution of Chaos in Living Matter

Discover the dazzling beauty of science in distinctly
new exciting ideas about life's origin

Farid Zapparov
Author of *the Food Delusion*

ISBN: 1490323422
ISBN 13: 9781490323428

Library of Congress Control Number: 2013910961
CreateSpace Independent Publishing Platform
North Charleston, South Carolina

Acknowledgment

I owe my deepest gratitude to Simon Webb for his help in editing and his sincere interest in the subject of the book.

Contents

"Some of us should venture to embark on a synthesis of facts and theories, albeit with second-hand and incomplete knowledge of some of them—and at the risk of making fools of ourselves." **Ervin Schrödinger, *What Is Life?***

Introduction

The systematic review of nutrition in terms of obesity and degenerative disease in my last book, *The Food Delusion* [1], showed the need to address the problem by an in-depth analysis of the evolution and origin of life. Based on the "viscoelastic prebiotic molecular evolution" hypothesis put forward in that book, there is a realization that the process associated with the primary origin of life is not a single isolated event. Rather, it was and still is continuously repeated greatly in all living system processes.

Most researchers interested in the problem of life focus on the origin of cellular life on the planet. However, life does not necessarily begin in the cell form. Before the process of the evolution of matter overcomes the distance between the reacting chemical soup and a state of a living cell, it needs some intermediate process. The state of matter in this process is no less alive than the state of matter in the form of a living cell. The arrangement of the simplest cell is incredibly complex and it consists of many very complicated supramolecular structures. The prevailing view is that there was a random formation of cells with phospholipid membranes. Within the cells chemicals somehow evolved into the states of supramolecular structures. Theoretically, it is possible.

This book focuses on the key question: How did the molecular evolution start and take place in the chemical environments in the cells or outside it? Also, what is the criterion for the chemical system to embark on along the path of evolution of living matter? I wish to draw on the very approximate outlines of processes involved and physicochemical pathways in the beginning of molecular evolution and how these processes manifest themselves in modern biological systems. Besides, we should understand that the theory of the origin of life is, in a sense, the theory of evolution. Based on the proposed items in the book, the properties of matter relevant to life could arise even in pre-cellular stage of evolution. Therefore, in my view, the approach of James Lovelock [2] to the biosphere in terms of the Gaia concept is the most reasonable. In the initial state of pre-cellular life on the planet is the evolution of Gaia. For Gaia, basically it is not so important what the life form is—cellular or noncellular—but the nature of the reactions in the process of molecular evolution.

Earth accumulates the energy radiated by the sun and heat. Then between the heated planet and space, with a temperature of only around four degrees Kelvin, a temperature gradient appears. Because Earth has a motile gas and liquid cover (air-sea), the temperature difference causes in them a convective chemical cycle. This cycle also involves the solid crust to a depth of dynamic erosion. This leads to global geochemical circulation—a necessary component of Gaia.

Despite the huge developments in physical and chemical science and applications in the fields of molecular biology during the past sixty or so years, the E. Schrödinger's question [3], "What is Life?"—unanswered—stubbornly continues to pursue academics in their endless discussions about the origin of life. Most ordinary people and even nonbelievers in God trust in the existence of an exceptional creative act for the origin of the first primitive cells. Moreover, even many biologists believe in it.

Life, as proposed in the point of view of this book, is a unified evolutionary process, including molecular evolution in the so-called chemical phase, expressed later in the Darwinian evolution. This process began with the chaotic reactions of organic compounds in the "primordial soup." Then after a certain physical process of renormalization inherent in matrix molecular systems, it has led to the emergence of the first biopolymer molecules and supramolecular structures based on them later. However, the process has not stopped. Through the process of chaotization and renormalization of the living molecular matrix, new states of the evolution of systems involved in the process begin and continue. This process of trillions of discrete acts continues every second up to the present day in all living systems.

Meanwhile, life is regarded by many scholars as a unique phenomenon in the form of a large long living fluctuation, which once arose on Earth or elsewhere in the cosmos. This is based on an intuitive view of the majority of people of the existence of the contradiction between the phenomenon of life and thermodynamics, although analysis and measurements show that living matter exists in perfect harmony with the laws of thermodynamics. Nevertheless, the idea of uniqueness significantly limits the ability of the scientific approach to the problem of the emergence of life. And as long as we do not destroy the idea of uniqueness, the problem of the origin of life, strictly speaking, is doomed to be the subject of philosophy, theology, and science fiction— just what you like—but not science. This is so because the most important hallmark of the scientific method is the capability of experiments to be reproduced. That is why most biologists discuss this issue with undisguised hostility.

While this is a significant problem for the proponents of the uniqueness of the origin of life, according to my conception, life continuously emerges from an inanimate state and returns to it again at the molecular level of the structure of any living organism. Moreover, the basis for the existence of macroscopic cellular life is an ongoing set of cycles of birth and death at the molecular level. Each discrete part of the process of destruction of the molecular structure—the transition to chaos—"is a little local death." However, the transition to a new state is a new local birth of life. The ratio of these processes determines whether we grow or age and their nature has not changed at least in the last four billion years.

But since life is not unique, then why have scientists, despite many attempts, not been able to create a new life using the chemicals found in nature? Perhaps because of the fact that in the reasoning of many aspects of life, they do not appreciate one very obvious and general quality of life—the mechanical viscoelastic properties of living systems at all levels of their organization. Proposed in the book, the viscoelastic approach leads to the conclusion that life is not some mystical phenomenon. Rather, it is an inevitable phenomenon like a "phase" transition of chaotic inanimate matter to a state of directed evolution. This happens if and when the physicochemical parameters of the "primordial soup," achieve some critical transition values. In this case, life is inevitable, just as the fact that water is condensed into a liquid at temperatures below 100 degrees Celsius and freezes below zero.

But on the other hand, we observe life in the complexity and diversity of forms much superior to what we see in inanimate nature. The living state of matter is perceived by us as a wonderful thing. Obviously, this is the basis of the religious approach to the world. Even today, many scientists believe in the existence of "vitalism." Indeed, the morphology of life is so amazing in its manifestations that the very idea of its dependence on the physical laws at the molecular level seems to many to be blasphemous. Therefore, finding the elements of a great miracle—living matter—is a major challenge of modern science. To attempt to describe the elementary act of evolution, which led to the emergence of life on our planet, that is, Gaia, is the subject of this book.

From molecular chaos emerged supramolecular structures. But another chaos—chaotic changes in the environment—caused continuous total or partial destruction of these structures. However, some property of the system, which initially contributed to the emergence of the first structure, continued to promote the formation of new structures and the selection of those resistant to given specific environmental conditions. What is breaking the supramolecular structures' force is more or less clear—it could even be the thermal motion of the molecules. But what was the property of the system that led to the supramolecular structures' formation, and what was the selective force that conducted the selection and provides a stable evolution of these structures?

Of course, it is understood that the emergence and existence of life in such a scenario is possible in a certain range of external parameters. This produces amazing macro effects when reacting systems behave fantastically in a commonsense manner. For the possibility of a living state to emerge, the "future" state of the reacting chemical system must determine its "present" state. The influence of the "future" on the "present" happens through the partial presence of the "future" in the immediate "past" of the "present" in the depth of time called the relaxation time. Does it look a bit complicated as time travel extravaganza? It is! But this is possible in viscoelastic systems.

That inversion of the natural order of time is a "miracle" that determines all other wonders of life through the process of the evolution of such systems. Of course, the concept of time and the future here have a certain sense, which does not allow natural laws to violate the principle of causality and cannot permit travelling through time. The appearance and the presence of the viscoelastic properties of living matter can explain such wonders through existing laws of physics without using the concepts of vitalism, or search for other new laws managing the universe aimed at ensuring the existence of living matter.

According to developments in the approach to the problem in this book and earlier in *The Food Delusion*, the necessary and sufficient condition for the occurrence of molecular evolution leading to life in the known forms and perhaps in the forms unknown to us is the emergence of the viscoelastic properties of the medium in the primordial soup. Suggested principles governing the emergence of life, theoretically, can solve some of the problems of chemical evolution that cause difficulties for scientists. The main idea of this book is that the emerging viscoelasticity stipulates the evolution of macromolecules and supramolecular structures. Still, the question remains of what superposition of the chemical compounds, relaxation times, and phase states in the process of molecular evolution are making the transition from inanimate systems to life. The answer to this question would require another set of studies. But, nevertheless, I hope that the proposed ideas arouse interest among professional scientists and in the general public.

I
Rheology of life

If you ask the question, what is the exclusive property of living matter that is uniquely related to its living state, then we have thermodynamic-type answers about negative entropy, dissipative structures, the exchange of energy and matter with the environment, the existence of limited cycles, etc., in their numerous variations by different authors. Without going into the details of all answers ever given to that question, I would argue that all of these features are in biological systems, but they are not exclusive to their descriptions. This assertion is based on the fact that they are also observed in inanimate nature.

The exceptional property of living systems is only their viscoelasticity. Viscoelasticity almost does not exist anywhere in inanimate nature. Of course, most synthetic polymers have viscoelastic properties. But first of all, polymers have been created by living systems—people—and never existed independently. Viscoelasticity is sometimes used in the modeling of tidal deformation of the planet's crust and the liquid mantle. However, this case is not directly related to the molecular level of the existence of life. But, nevertheless, the presence of a planetary tectonics is considered by many scholars a very important condition for the development of life. It is because the tectonically active planet delivers minerals and gases necessary for the existence of life through the fissures or volcanoes to the surface.

Viscoelasticity not only exists exclusively in living beings, but all living beings possess tissue viscoelasticity. Viscoelastic mechanical properties are represented at all levels of the hierarchical structure of organisms, ranging from biopolymers DNA in the nucleus to all over the body. The viscoelastic properties of matter are studied in rheology—the science of deformation and flow of continuous media, which has the elastic and viscous qualities simultaneously.

I must note here that as a scientific discipline, rheology is considered as dull and boring part of all sciences. It is the antithesis of popular public scientific fields such as, for example, the theory of dark matter, cosmological wormholes, molecular biology, the theory of evolution or the Big Bang, etc. Most scholars versed in the mechanics of continuous media say that the unique rheological properties of polymers trivial are a consequence of the length of their molecules. Engineers and technologists in the field of plastics or rubber processing consider rheology as a very unfortunate misunderstanding and a mistake of nature. Nine out of ten biologists and physicians, ten out of ten economists, lawyers and geographers, and half of all other scientists are unaware of the existence of such science as rheology. Rheology is very close to the other prestigious science—proctology—because the motion of shit is the object of both disciplines. Along with that, it solves other exciting tasks, such as the dynamics of fecal matter in sewage. What a fascinating science!

Experts in the field of fluid mechanics and the dynamics of continuous media are well aware about this science and hate it for its failure to make reliable calculations in engineering problems. Solutions of mathematical calculations in rheology are obtained, basically, only by substituting the experimental results, i.e., substituting the answer in the solution itself. This creates in some of the most conscientious scientists and engineers a huge inferiority complex.

Still, rheologists, overwhelmed by all the problems, with hypocritical sympathy from more fortunate colleagues in other branches of science with barely concealed contempt, are hiding from the rest of the world the charm and beauty of their science, which almost no one, apart from themselves, can assess. The fact is that most of rheology tasks are hardly solvable problems with "super nonlinearity." This means that the nonlinearity of the processes "encounters" the nonlinearity of the medium. This gives rise to qualitatively new amazing effects that you will not find anywhere else in nature. One of the most striking effects of the intersection of these two forms of nonlinearity is the "magic" of life, which allows us to assess the immense importance of rheology.

Therefore, in this book, I, as a former rheologist, will try rehabilitate my colleagues to the world, proving that the rheological properties of matter in the subject of the origin of life and evolution are essential. If we take the standard issue of the chicken and the egg, the rheological properties of molecular environments of living matter is the most fundamental, that is, primary in respect to the molecular coding in DNA and cellular structure of all organisms. The presence of the rheological properties of living matter determines the diversity of its current evolutionary morphology and the very origin of genetic code. Everything in the world, from the beauty of butterflies or flowers to the mind, the glory of the sky and the ocean, love and children—all that is beautiful on Earth arose due to the special rheological properties of polymer molecules.

Additionally, it should be mentioned that the most inquisitive, interested mankind have two large explosions that occurred in the history of the universe: the first, called "big bang," happened about fourteen billion years ago; and the second, which happened about four billion years ago, marked the emergence of life on Earth. Two almost identical main hypotheses exist about the nature of each: the first is about Almighty God in different forms; the second, which gained wide publicity in the last one hundred to two hundred years, is the hypothesis that both of these phenomena are fundamentally explainable in terms of physical laws governing the universe. From my point of view, an attempt to make a small but critically important contribution to the second hypothesis is the subject of this book.

2
Viscoelasticity

Let us briefly examine the ability of some media to have elastic and viscous properties simultaneously. Elastic deformations occur in the body when a load is applied and disappear when the load is taken off. Viscous flow occurs at any arbitrarily low tension. With growth of tension, flow rate increases, while maintaining the stress, and the viscous (liquid) flow continues indefinitely. When matter is resilient and fluid, it is called elastoviscous liquid. Typically, this property of living matter is in the form of biopolymers. Another property that polymers may have is high elasticity, typical of materials such as rubber, which allows multiple stretching, and after removal of the load almost instantly recovers to its original state. These two properties can be combined with one common term: viscoelasticity.

The typical rheological process is a relatively slow flow of matter in which are found viscous and elastic properties. Rheological phenomena appear in many natural processes, of which we are only interested in the processes of living systems. Fluids with rheological properties are called non-Newtonian fluids. Effects arise on flow of viscoelastic non-Newtonian fluids, which are quite peculiar, for example, the Weissenberg effect, where a polymer solution of high molecular weight rises up around a rotating vertical rod. The other variant of this effect can be shown as two identical glasses—in the left (Fig. 1), a Newtonian viscous liquid, and in the right glass the polymer solution of the same viscosity; each cup is rotated around its axis. Stationary round bars are lowered into the glasses from above. In the glass with the ordinary liquid, we can see a predictable pattern— a flow takes the shape of a revolving body with a maximum height of parabolic surface at the wall.

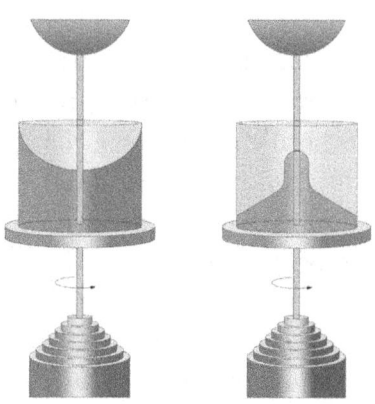

Fig. 1

Another interesting experiment is with solutions of polyoxyethylene or polyacrylamide in water. On pouring the polymer solution from an upper glass into a lower glass (Fig. 2), we find that the flow of the solution continues from the upper to the lower glass, even if positioned as shown in the figure. A liquid jet rises up the vertical wall of the upper glass, and then overflows and runs down to the bottom glass.

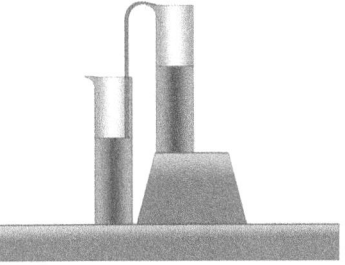

Fig. 2

A mathematical model of the rheological properties of a viscoelastic medium is defined by the equations relating to the stresses and strains and their first-time derivatives. This equation is called the rheological equation of state of the medium. Theoretically, on the basis of this equation, we can calculate the stresses and strains in the flow. However, in practice, the equation of state reflects the real properties of the medium very roughly. This is due to the incredible complexity and diversity of polymer media and dependence of their properties on a variety of parameters.

No less interesting is the effect of significant reduction in the hydraulic resistance in turbulent flow of low molecular weight liquids with small additions of (sometimes a few parts per million by weight) polymers. It turned out that we can achieve 75 percent reduction in the hydraulic resistance of water in a pipe. This effect is frankly used very little in a few applications in industry. In the recent past there were attempts to apply this effect to accelerate the movement of military boats in the fleet. Furthermore, the viscosity of viscoelastic polymers depends on the flow shear rate. Usually in polymers with increasing shear rate, the viscosity of the solution is thinning down, while the viscoelastic fluid moving in the channels at high shear rates is observed to have pulsations and other phenomena that are specific to non-Newtonian flows.

3
Polymers

Viscoelasticity and the rate of polymerization are highly, although ambiguously, related to each other. The ambiguity stems from the fact that one and the same polymer may have different states of its macromolecules. And then the viscoelastic mechanical properties of the polymer in these states are different. A molecule of any substance has a definite arrangement of atoms in space, or, as they say, have a chemical structure, conformation, and configuration. Atoms and atomic groups in molecules are not fixed and are able to make a variety of movements relative to each other, oscillations, and rotation. As a result, these movements change the length of the chemical bonds and the magnitude of bond angles (the angle between the directions of chemical bonds), but in a small range.

As well, even low-rate polymerized molecules can be arranged in a matrix with high viscoelasticity. Polymers are long molecules with a high molecular weight. For small molecule chemicals, the molecule is the smallest particle of matter that cannot be crushed without losing the basic chemical properties of the substance. Virtually any polymer is a mixture of macromolecules with different molecular weights. The length of the macromolecules of the polymer or the degree of polymerization is defined by the number of monomer units in the polymer chain. Relatively short molecules are called oligomers. For example, if a macromolecule of polymer can contain millions of units, the oligomer molecule is usually limited to from tens to thousands.

But it is known that for a lot of polymers, the terms "molecule" and "molecular weight" have lost their conventional meaning. There are substances that have molecular matrix with chemical or physical nodes. In the physical chemistry of polymers they are called gels. A chemically cross-linked gel literally is one giant macromolecule. The term "molecular weight" cannot be applied to such substances. For gels, we can talk only about the spatial density of the matrix and the molecular weight of a piece of the chain between neighboring nodes.

In polymers, solutions usually dominate the physical connection between the sections of the macromolecules, which are relatively easy to break down. Therefore, chemically cross-linked gels generally have a yield strength due to reformation of chemical bonds of the matrix by applying a strain. Polymer solutions or gels, in which the physical connection may be easily broken, have no appreciable yield. While the molecular weight of the polymer is relatively small, all physical and chemical properties of the molecule are changing rapidly with rise of molecular weight. However, after a certain limit, further increase in the molecular weight is no longer a significant factor affecting polymer properties. This limit is reached most quickly for aggregate states (some liquid oligomers' turn to the "glasses" is the degree of polymerization at twenty to thirty).

4
Relaxation

Relaxation of stress or strain (creep) is an essential characteristic of all polymers. Relaxation is a process of statistical establishment of equilibrium in the system. The rate of establishment of equilibrium is related to the probability of transition from one equilibrium state to another. Causes of these transitions are different: change in the oscillation frequency of the particles (segments of macromolecules) making up the substance, rearrangements of these particles and their aggregates, changes in the macromolecular conformation or entanglement network, etc. The rate of establishment of equilibrium is characterized by the relaxation time, which is exponentially dependent on temperature.

The specific feature of polymers is that they have primary structural elements (links, segments, and chains) combined in a variety of supramolecular structures resembling the structure of crystals. The interaction of the solvent with the polymer macromolecules in solution by hydrogen and other types of bonds also influences the formation of supramolecular structures. Therefore, in the polymer solutions there may be a movement of a variety of the structural elements that define a set of relaxation times in such a system. The rate of relaxation is actually dependent on a myriad of factors, including the magnitude of the potential barriers of states of elements, temperature, environment, etc.

The relaxation times of the structural elements of polymers can change in a very wide range from 10-10 seconds to 1010 seconds, and the corresponding relaxation processes are observed during loading or deformation of polymers by dynamic or static methods.

Any system relaxes according to its inner law, irrespective of the mode of excitation. During mechanical relaxation, the system is derived from the equilibrium of mechanical forces. At structural relaxation, the system does not depend on an external force and it moves to the equilibrium state, for example, by changing the temperature, since at each temperature the polymer has its own equilibrium. It follows that the simple, individual relaxation processes related to the mobility of the structural elements of this subsystem should be the same, regardless of whether we are measuring the relaxation times of structural or mechanical methods.

5
Phase states

The description of the states of matter is distinguished aggregate and phase states. Physical states (solid, liquid, gaseous) are different by intensity of thermal motion of atoms and molecules and their packing density. Phase states of matter differ in the order of the particles arrangement and thermodynamic properties. There are also crystalline and amorphous phase states. The crystalline phase is characterized by long-range order in the arrangement of particles and amorphous—the lack thereof. Long-range order refers to the order existing in the range, being many times greater than the distance between the particles. The particles of crystalline substances form a spatial crystal matrix, with atoms (e.g., carbon in diamond) or molecules (e.g., water molecules in ice crystals) in the nodes.

The concepts of phase and the physical state of matter in the description of the state of polymers do not always coincide with their analogies for substances with low molecular weight. Indeed, in contrast to low-molecular-weight compounds, where atoms and molecules are kinetic elements, the kinetic elements in polymers are much more diverse. It may be atoms or side atomic groups, monomer units, segments of macromolecules of various lengths, whole macromolecules, supramolecular structures of different sizes, and different structures.

Under the term "supramolecular structure" I mean the way of packaging macromolecules, their relative position to form spatially specific structures with a certain degree of internal order, and the character of intermolecular interactions.

In thermal motion in the amorphous polymers, as mentioned above, are involved different kinetic units. Depending on which ones are dominant in the amorphous polymers, there are two physical states existing: glassy and viscoelastic.

Strictly speaking, the viscous and elastic state of the medium is characterized by two types of behavior: first, viscoelastic behavior is typical of chemically cross-linked gels; second, elastoviscous behavior is typical of polymer solutions and melts. Polymers in this state are able to flow under the influence of an applied stress.

In the viscoelastic state, thanks to chemical cross-linking, it is impossible to move the macromolecules as separate kinetic units. Viscoelasticity is possible only at a low frequency of cross-links, when the distance between adjacent cross-links is bigger than the chain segment.

Polymers in this state have amazing mechanical properties: they are able to experience enormous reversible deformation, sometimes reaching several hundred percent. However, sometimes the bridges between macromolecules have the character of physical links or the mesh of chemical bonds is easily breakable and easy to restore. In this case, the system is a gel. Starting from a certain tensile strength, it is still possible to limit the flow of such a system.

Despite the fact that these two viscoelastic and elastoviscous states are different from each other, in what follows, for simplicity I will use the term viscoelastic for both states.

6
Gels

Organisms of all living things are made up of an infinite number of biopolymer matrices forming gels. In principle, the human body can be considered as one large heterogeneous gel. Therefore, let us take a closer look at the state of matter as a viscoelastic gel. Gels (or jellies) are polymer matrices in a state of macromolecular concentration higher than the percolation threshold (percolation represents the flow of fluids through porous media or matrix) [4]. The nodes in the gels can have both a chemical and physical nature. Stable gels behave in a certain range of loads as deformable solids. The inability of gels to flow is associated not only with the presence of the matrix mesh itself but also with the immobilization of the solvent in the matrix: even a very dilute polymer gel often could be cut with a knife. Immobilization of the solvent happens at the elementary matrix mesh level. Inside these matrix meshes diffusion of the solvent molecules takes place almost as easily as in a solution of equivalent concentration.

Gels capable under mechanical or chemical influences go from one state to another by changing the volume or breaking up with a "collapse" of the gel. Sometimes decay is not into two phases, with high and low concentration, as the pure solvent and the concentrated gel. It looks like a solvent is squeezing from the matrix and is called syneresis. Syneresis may be caused by changes in temperature or concentration, and sometimes occurs spontaneously. This spontaneity means simply that the original gel was in the thermodynamic nonequilibrium state or in other words «on the edge of collapse». But for some kinetic reasons (immobilization of the solvent, the presence of interference of mesh lattices, high viscosity) it could not move quickly enough to the equilibrium state. The transition of the solution into a gel is called sol-gel transition, and the reverse is gel-sol transition. An important characteristic of the gel matrix is the degree of polymerization of the linear chain segments between nodes. The significant difference of physical gels from simple chemical gels is that for the last degree of polymerization—constant. In physical gels, depending on the nature of the nodes, in reformatting matrix, it can be quite varied.

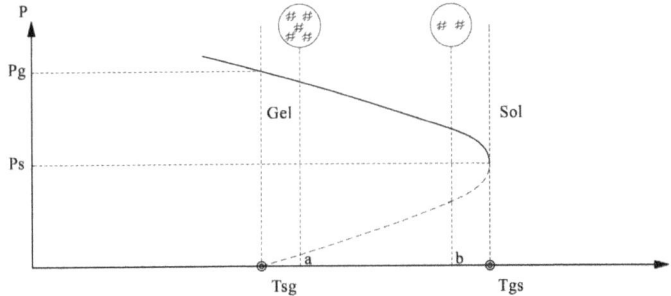

Fig. 3 schematically shows the state diagram for sol-gel and gel-sol transitions, when gelation temperature Tsg state transition occurs from solution into gel state. However, the reverse transition for biogels typically occurs at a higher temperature, i.e., hysteretic type. Points **a** and **b** show the change of the number of the gel micro blocks (Chapter 8) according to the distance from the point of sol-gel transition.

Gels are thermodynamically stable away from the phase transition region. However, their kinetic stability is determined by the structural relaxation time of the nodes. Accordingly, when the gel is exposed to the periodic forces at a frequency considerably higher than relaxation time in inverse power, it will behave as elastic material. When the frequencies are considerably lower, the gel will act as a concentrated polymer solution, where only the physical nodes exist. That is, at certain frequencies of mechanical influence on the gel matrix, it will skip the mechanical waves through itself as an elastic body. At the lower frequencies, large energy dissipation in a viscous solvent phase will inhibit the propagation of waves.

Despite the presence of the matrix, polymer gels are a typical two-component system. Under the action of some external power, nodes can be partially destroyed, and the gel becomes fluid. Since the restoration of sites require some time, the solution remains liquid and only gradually restores the properties of the gel. This phenomenon is well known as thixotropic behavior.

If the nodes are not the chemical nature (although in many cases it is possible on occasion to break the weak chemical bonds) but have a supramolecular structure, with a certain tensile strength, they can start to break or move. In turn, this leads to the fact that the net is temporarily transformed into a chaotic transient state, culminating in the restructuring to a new state.

The transition from a state of chaos to the ordered state and the reverse transition in gels play a big role. This role is particularly significant for heterogeneous gels, where plots of polymer between neighboring nodes differ in physical and chemical composition. The transition to a new state of the individual chain influences changes in the rest of the matrix. However, the complete collapse of the gel, as

is usually the case in homogeneous gels does not occur—
the matrix is rebuilt to the new state. The wave of strains
is propagated throughout the gel matrix and initiates the
transformation of the entire matrix of the gel. The transformed matrix changes the conditions of molecular interactions that could theoretically occur in the gel matrix. Such
change in the system state is called bifurcation. Local decay (chaotization) and the restructuring of supramolecular
structures initiate in a 3-D polymer matrix the propagation
of changing state waves.

7

Singularity of bifurcations and black holes

Biological systems display nonlinear behaviors that I have described in *The Food Delusion* [1] and can be widely observed in the natural world in many other occasions. Their state variables can change discontinuously, producing fast and sudden changes of state—bifurcation, which also may be called phase transitions. Examples of such phenomenon are a superfluid, superconductive, ferromagnetic, and, as I am trying to show below, the viscoelastic states of matter. In transition to such states, some values of variables (pressure, temperature, concentration, etc.) often play a decisive role. A critical phenomenon exists where a dependent variable initially does not interact or interacts very weakly to continuous changes as an independent variable, but beyond a given point they may start to interact. The "limit" may be defined by a critical concentration, temperature, etc., for example, in systems with chemical reactions. The nonlinear dynamic systems in conditions above the "limit" with increasing relevant variables begin to move spontaneously from chaos to a new orderly stationary state. Furthermore, systems characterized by strong nonlinear dynamics can display a specific kind of irreversibility, called hysteresis. This means the path dependency of phase transitions is in terms of bifurcation: a point where a trajectory can proceed in different directions. By the way, in one study [5] at the generalization of Prigogine's theorem for nonlinear systems far from equilibrium, the influence of processes of relaxation on physicochemical processes has been studied. It was shown that the phase transitions of the second type—analogous to "jumping" bifurcations, which emerges and fades out by hysteresis trajectory—exists in such systems. Bifurcation in this case refers to a qualitative change of the system in a continuous monotonic variation of a system parameter, called the bifurcation parameter. With ever-changing parameters, the sequence of bifurcations can occur.

The bifurcation of the type under consideration may illustratively compare to the passage through a barrier, in which there is a hole equipped with a valve opening in one direction only. The return to the original state, if possible, happens only through the hysteresis loop, i.e., through another hole and valve, or, strictly speaking, on the other path. That is, there are two points or two set parameters in which a bifurcation can happen, depending on the history of the system before the transition. Therefore, under continuous change of the parameter, the direction of the parameter's change in the hysteresis region of bifurcations is an important factor.

Hysteresis is a property of any system, expressed in the fact that it's internal characteristic of nonlinearly is dependent on external factors. Functions that define the state of an object depend on the history of the force exposure on the system. That is, hysteresis exists in the systems that have a memory, as viscoelastic systems. Hysteresis occurs because the change of stresses or strains in such a system always takes some time, called the relaxation time of the system.

Of course, all processes in nature take time. However, there are fairly large biopolymer relaxation times, from a fraction of a second to hours.

The slower the change of the external characteristics influencing the system, the less time is needed for internal characteristics to adapt. In some of the viscoelastic processes, because of the existence of the relaxation time when loading and unloading the system, they go through different microstates. Then it is said that the forward and backward trajectories of the system in phase space is different. This behavior is called hysteresis. Thus, with viscoelasticity, the elastic deformation completely vanishes when the load is removed, but with a delay determined by the relaxation time of the system. Hysteretic type of phase transition, as will be shown below, is one of the most important factors determining the characteristics of self-organization in biological and prebiotic structures.

In going through the bifurcation point from a higher to a lower level, one might result in the degeneracy or an annihilation of order parameters of the higher level. In this case, only two modes of system state are possible: stable stationary point (an annihilation of order) and a limit cycle. A. T. Winfree [79] called the area between the points of bifurcation, in which there are two modes (two strange attractors), a stable stationary point and a limit cycle, Black Hole. By analogy with the cosmological black holes, when a system passes the area between these two bifurcation points, information about the ordered state of the system at a higher level in the hierarchy is disappearing. But of course the term "Black Hole" as used here in context of phase transition between different orderly states should not be confused with the same term in cosmology. But the similarity noted by Winfree is striking.

In this range of parameters, theoretically, if we can somehow damp the perturbations of finite amplitude, we may arrange then that the system falls into the basin of attraction of a stationary point, which will lead to the disappearance of the structure determined by a higher level of the system hierarchy. In particular, the problem for the Benard convection in non-Newtonian fluids [7], this will lead to a damping of convection; or in the gel on the edge of collapse to the disappearance of the gel phase [4], etc.

Because viscoelastic materials can have the viscous behavior of a fluid and the elastic behavior of a solid simultaneously, there are many features of their mechanical behavior. The presence of relaxation processes leads to a certain lag in the feedback according to their characteristic relaxation time. For example, according to [6] the network of protein molecules in the cytoplasm of the cell, changes in the state of one macromolecule affects the state adjacent to some characteristic time lag and so on.

With that, at movement of the signal of altered state through the network in each molecular section, this lag time varies according to the characteristic relaxation time for each intermediate molecule. The same applies to the mechanical and chemical feedback loops from changes in the state of macromolecules in a network of proteins. The influence of non-Newtonian or viscoelastic properties manifests itself in additional nonlinear processes. In [7, 8], the simplest dissipative system is a process of Benard convection in thin layers in a nonlinear viscoelastic fluid of the second order. Such fluids have been experimentally simulated by polyoxyethylene solutions (a polymer with significant non-Newtonian properties). The evolution of the wave packet disturbances of finite size at the bifurcation point of heat transfer in a nonlinear viscoelastic medium has been studied. It was found, theoretically, that the bifurcation responsible for the process onset of convection, and attenuation due to the equation of liquid state additional nonlinearity, has hysteresis (Fig. 4).

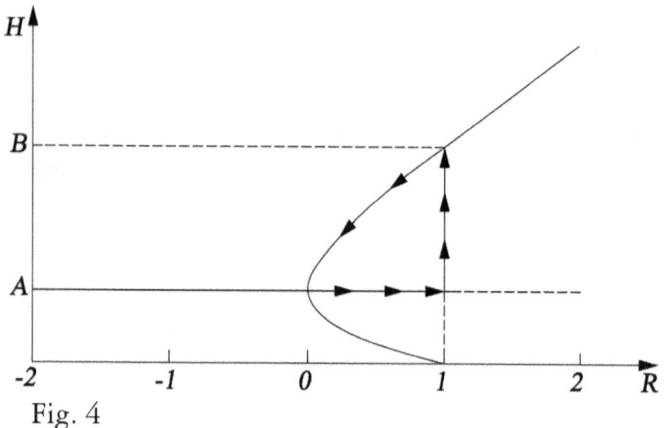

Fig. 4

Both transitions occur at different normalized parameter R. In [7] case, normalized temperature drops (Rayleigh numbers). The phenomenon of excitation of convection at point (1, B), in fact, is a bifurcation of the process of heat transfer by thermal conductivity at low temperature differences, if there is a simple dissipative structure. But convection in nonliner viscoelastic liquid will stop at point (0, A).

Consider illustratively what will happen in a nonlinear viscoelastic fluid if we move along in the parameter R, since negative values (Fig. 4). For R <0 steady-state solution, a stable limit cycle. For 0 <R <1, the two solutions, an unstable and stable limit cycle. Initially, there is a single stable steady state to R = 0, all perturbations decay. For 0 <R <1 there is a stable limit cycle, thermal conductivity of the stationary state (infinitely small perturbations decay). However, once R is positive, the steady state becomes unstable, and R = 1 is a sharp jump from point (1, A) to a stable limit cycle to point (1, B). The system exhibits disturbances of finite amplitude, the onset of convection. If we move from positive to negative values of R, the perturbations of finite amplitude exist as long as R becomes less than 0, and then disappear. Thus, for $0 < R < 1$ there can be two different states. Which of them is realized depends on the history of the system. This phenomenon is called the effect of the viscoelastic hysteresis. Then, of course, similar hysteresis processes can occur in other more complex dissipative structures. And in particular, it appears to be in the flow of physical and chemical reactions of metabolism in the body's viscoelastic media.

For example, in the gel-like environments of living organisms, the elastic biopolymer matrix frame supports a stable structure. Physicochemical reactions in the low-viscosity phase of the gel are accompanied by the movement of groups of atoms. The low-viscosity phase, where the conformational (shape of the molecules) relaxation of matrix elements occurs, is not an insurmountable obstacle to the movement of some atomic groups of biopolymers in a certain range, while undergoing a physical and chemical reaction. The conformational change of biopolymer matrix elements will affect the elastic properties of these elements and the matrix as a whole. But the movement of atomic groups, whose geometry does not fit into the limits of the conformational changes at a given stress, will be difficult. Physical and chemical conditions for conformational and thus viscoelastic specificity of chemical reactions in the space of a biopolymer matrix are thereby created. From a general point of view, the cause of this specificity is that the redistribution of chemical bonds during the reaction changes the electron density distribution in the reacting molecules, thereby changing the balance of forces (stresses) within the macromolecules, which leads to the fact that their conformation before and after the reaction are different. Often such a conformational change is reversible. The reverse transition at the termination of the reaction can be considered as a conformational viscoelastic relaxation of uncompensated stresses in biopolymer macromolecules. An example of such a direct conformational transition is the addition of oxygen to hemoglobin molecules. Joining a very small molecule of oxygen to hemoglobin causes significant changes in the conformation of certain proteins—the rotation of some atomic groups of molecules at angles up the order to ten degrees.

Any change in the chemical characteristics of the environment in which the macromolecule abides alters its conformation. Also, mechanical deformation of the macromolecule, equivalent in magnitude to that caused by the changing environment, causes the same change in its chemical potential. Chemical potential is the amount of the internal energy in the system, which can be spent on chemical transformations per unit mass of the medium. That is, change in the conformation of macromolecules can be caused by changes in the chemical potential of the medium in which they exist and vice versa, a mechanical change in the conformation changes the chemical potential of the macromolecules themselves.

The main macroscopic effect of polymerization manifests itself in the viscoelasticity. But can we account the viscoelastic state as a super-state like the super fluidity in helium-H, superconductivity, ferroelectricity, and ferromagnetism? I think at the level of mechanical properties, there is sufficient reason to consider viscoelasticity as a super-state because its appearance is associated with a phase transition, polymerization. Moreover, according to the concept of this book, the transition from the low molecular weight with Newtonian rheology of "primordial soup" into the viscoelastic state is the basis for the emergence of life.

From the point of view of the immense importance of this transition for the emergence, the type of matter that classifies different phase states, i.e., you and me, dear reader, this transition undoubtedly deserves classification as the transition to super-state.

The hysteresis type of excitation and attenuation of reactions or phase transitions in turn affects the character of the influence of feedback loops, the main function of which in a self-organizing system is to control the concentration of certain types of particles, the chemicals in a particular place (renormalization). That is, there is an additional nonlinearity. The existence of nonlinear viscoelasticity is an important element that affects the processes of metabolism (chemical changes in the body) and, apparently, is responsible for their stability at all hierarchical levels of the body.

Nonlinear systems display a number of behaviors that can be widely observed in the natural world. Their state variables can change discontinuously, producing phase jumps, i.e., sudden changes of state, as in the phenomenon of ferromagnetism or in superconductivity. In such discontinuous processes, threshold and critical mass phenomena often play a decisive role. A threshold phenomenon exists where a dependent variable initially does not react at all, or only very little, to continuous changes of an independent variable, but beyond a given point it reacts suddenly and strongly. The threshold may be defined by a critical mass, e.g., the number of particles of a specific kind that must be present before a reaction sets in, but other kinds of threshold also exist. There are phase transitions from order to disorder and in the reverse direction. The behavior of nonlinear systems can become completely irregular—or "chaotic"—if the values of given parameters move into a particular range. On the other hand, nonlinear systems can also move spontaneously from disorder to order, a stationary state far from equilibrium; this is called a self-organization. Furthermore, systems characterized by nonlinear dynamics can display a special type of irreversibility, i.e., hysteresis (path dependency of phase jumps), and a specific kind of indeterminateness expressed in the term bifurcation: a point where a trajectory can proceed in different directions. Existence of the phenomenon of self-organization is entirely determined by the presence of the phase transition hysteresis.

As proof of this discussion, let us look at the theoretical approach of one study [9, 10], which shows that in the physical-chemical systems, if the instability to finite perturbations (bifurcation) appears-disappears (direct-inverse transition) according to the mechanism of the hysteresis type (viscoelastic systems), then the macro states of the system determine its micro states. The situation is in a way the opposite of what exists in the classical linear medium, where the ensemble of micro states determines the macro states of systems. For the viscoelastic systems, the macro state provides the realization of corresponding microstates through renormalization of probabilities, as a result of which the ensemble of possible micro states narrows [9, 10]. Such a renormalization of probabilities means that the probabilities of the states of the ensembles of particles that do not lead to the macro state of the system become negligible. From the physical and chemical point of view, this means that not all the possible physicochemical configurations of molecules can participate in the processes of the structure's emergence. In this way, the viscoelastic properties of a medium determine the main feedback mechanism, which in turn determines its properties on the micro level. They also determine the variety of emerging structures and their stability due to having quite big relaxation times.

8

Selective factor of evolution

We touch on more of the process of renormalization, which, in other branches of physics, many perceive as a mathematical trick, due to the imperfection of the mathematical description of processes in a particular area of physics. The renormalization approach emerges from the necessity of renormalization of the quantum variables in order to address the problem of infinities in a quantum field theory. In general, in physics the renormalization should establish a relationship between parameters of system in the large and small distance scales. The physical meaning of the renormalization process in the systems of polymer gels can be illustrated on the basis proposed by L. P. Kadanoff [11], the "block-spin" renormalization approach. This approach was further developed by the work [12] of Kenneth Wilson. Wilson's ideas had demonstrated the importance of the renormalization solution in the theory of phase transitions and critical phenomena. He was awarded the Nobel Prize for his decisive contributions.

The basic idea of the Renormalization Group theory is to study the variation in the importance of certain properties at a smaller or larger scale. But this method also may apply to some physicochemical discontinuous processes, where threshold and critical mass phenomena could play a decisive role. In particle physics, it is applied through the concept of lattice gauge theory. The main idea of lattice gauge theory is that the emerging "infinities" (instead of the convergence of the integration results to the finite values) that appeared to be flaws in quantum field theory are actually the data you need to move from one scale to another.

In the gel forming system, the wave functions of the particle state, in the hysteresis loop, as it approaches to the gel point, tend to identify with the future function of the state of gel phase. When this is happening, the particle state function in the hysteresis loop is degenerate in some parameters. These are the physicochemical parameters of gelation. As in lattice gauge theory, in the gelation process, data appears (degenerated parameters), which defines the structure of the next supramolecular hierarchical level (gel). That gel level is the next structure in relation to the low-molecular chaotic solution of monomers. It is interesting to notice the existence of a very strange similarity between the physicochemical processes of gelation and quantum theory. Perhaps this tells us about the deep connection with each other and the universality of processes at different hierarchical levels of matter. And a connective tissue of the uniformity of processes is likely to be a nonlinear theory of system pathways dispersion in dependence of the perturbation scale.

When applied to our system, it is a way to define the final steady state of the whole system as aggregates of small-scale components emerging in a parametric region where the bifurcation hysteresis loop exists—it is the "block" idea. It is well known that for gels with the hysteresis type of phase transitions (bifurcations) sol-gel-sol in the parameters range within the hysteresis loop, at the approach to the transition point sol-gel, the system is having microscopic regions of gelation.

That is, in the bifurcation hysteresis loop, the system forms numerous dynamic microstructures, "blocks" of the future states of the system. These microscopic blocks of gel are unstable and so periodically arise and fall (disappear) again. We can say that most of these virtual prototypes of future systems are killed in the competition. The longest-surviving blocks are those that are the most adapted to the currently existing environmental conditions. In the circumstances of arising viscoelasticity of the system, it turns out that the most long-lived microstructure is the micro-structure with the highest relaxation time. Thus, the viscoelasticity of the system becomes a crucial factor of the selection in such an evolutionary process.

As a result of the phase space "filling" by the surviving "blocks," the system enters into a new phase in a qualitatively higher organized state. The system now behaves more orderly than prior to the phase transition point. Physically, it looks like the sol-gel transition in the system or the formation of other supramolecular structures. It was observed in many experimental studies that the sol-gel transition temperatures in most bio-gel systems were much lower than the gel-sol transition temperatures based on the hysteresis of the phase transition loops. The energy requirement is less for the gel-sol transition than for the sol-gel transition.

It should also be noted that the system within the bifurcation hysteresis loop can be represented as a strange attractor, which is a mathematical image used to describe nonlinear systems. Thus, the strange attractor is a segment of the evolutionary path from one point of the bifurcation to another. The strange attractor sort of attracts like a magnet with many different paths of the system defined by different initial values of the parameters. A very important role is played by cooperative, collaborative processes of interaction of particles in phase space, which are based on coherent (normalized by final macrostate) behavior of an ensemble of particle microstates of an emerging "future" structure. In this book, from different viewpoints, I will repeatedly go back to the renormalization procedure in viscoelastic hysteresis to describe its meaning from the point of view of statistical physics, nonlinear mechanics, topology, and common sense. So my repeated reference to this problem is related only to its exceptional importance for understanding the processes of the emergence of life and its current continuation. I also want to be sure that the biologists and the general public, most of whom are not very familiar with the above-mentioned branches of science, realize that this is not a mathematical trick, but one of the basic properties of nature.

The emergence of the above-mentioned gel blocks means renormalization of the ensemble probability of the particle state in the sol-phase before the phase transition. This is because an ensemble of probabilities of particle states in the "blocks" is much narrower and resembles the ensemble of the gel phase. Hence from the general ensemble of states of particles in sol-phase is the "cut out" part of the ensemble states. The closer the system parameters are to the transition border, the more the states that cannot be cut out in the gel phase. This is the physical meaning of renormalization of the ensemble probability state of the particles. In polymerization or gelation, renormalization causes a new phase, the polymer matrix, in which processes are described, mathematically speaking, as a completely different set of equations compared with those at the low-molecular state of the system before the phase transition. That is, singularity appears similarly to the quantum-mechanical breaking of scale symmetry in a field theory. The physical existence of such singularities in molecular evolution, leading to phase transitions, is the justification of the existence of renormal-ization as a physical process.

People not associated with the natural sciences are usu-ally instinctively afraid of terms such as a phase transition, or, especially, renormalization. But, in fact, these terms are not difficult to understand. I will try to explain. For example, the term renormalization, which seems incred-ibly complex and is derived from quantum mechanics, in fact can be observed in everyday life. Imagine that you are standing in the crowd at the entrance to a night club with a certain dress code. You will see a bouncer or a guard that lets some people in but not others. Thus, the requirements of the club, by imposing restrictions on the crowd seeking to enter the club, will narrow the number of people admit-ted to the club. This is the renormalization of the ensemble of people in the crowd by their "future" state in the club.

Now imagine that some visitors of the club come out from time to time to get some air. They are well-dressed, slightly drunk, and their young ladies look readily avail-able—there is a new phase state of the crowd at the club. This means that the "future state" of wishing to go into the club will be partially present in the crowd, for most of whom a visit to the club is the new phase state in which they have not been, at least, that night in the club. They have not been in it, but they know something about it. Most likely its "future state" will increase the desire of the crowd to get in, i.e., will be a directed evolution ensemble of states of the crowd in a certain direction.

It is also easy to find an analogy to such abstruse con-cept as «breaking of scale symmetry». Assume that entry into the club costs $100. That means those in the crowd who do not have $100 will not be able to visit the club. This means that the symmetry (equality of people in the crowd) is broken. It is also obvious that symmetry is broken because the club's visitors can be two sexes (and these days perhaps more). The aspirations and behavior of the sexes in the crowd are different, and, therefore, an example can help in the modeling of very complex asymmetric phenomena in physics. So the devil is not as black as he is painted.

9
Strange time dynamic of chaos

Because the body is nothing but interacting molecules (including the atoms and electrons of course) then it is interesting to see how the extremely complex metabolic condition arises and exists. In quantum mechanics, the Uncertainty Principle prohibits accuracy. Therefore, the initial situation of a complex system cannot be accurately determined, and the evolution of a complex system can therefore not be accurately predicted. Speaking of polymer gels, when the sol-gel transition occurs, it is a statistically selected state of the system, which "stretches" its properties in the domain of the system parameters between the points of gel-sol and sol-gel transition. Ensemble probabilities states of particles in this range of parameters are defined by the renormalization of the probabilities of the particle states in the gel phase. As it moves on the scale of parameter (e.g., temperature) from the gel-sol transition point to sol-gel point, the function of ensemble of probabilities of particle states narrows and turns into the function of the gel phase state at the point of gel formation. However, the system is chaotic in the hysteresis loop and is in so-called deterministic chaos.

The fact that in the gel phase not all states of the particles involved in the formation of a gel matrix are allowed is rather obvious. But the question arises, how does the system, when moving in the area of the hysteresis loop of gel formation bifurcations, know about its condition in the gel phase? Why does it tend to it?

The fact is that this is one of the main and amazing features of polymer systems that distinguish them from other states of matter. This feature of the polymer systems is the basis of the origin of life and its existence and evolution. According to the approach of Russian biologist G.E. Mikhailovsky [9, 10], to biological systems (populations), this property can be interpreted as the fact that the "future" macrostate of a system determines its microstate in the present. As it was shown above, this can be attributed to the biopolymer systems. But, in spite of the fact that it sounds like a statement about time travel, there's nothing mystical. The difference between polymer systems movement trajectories in the phase space near the bifurcation points determines the possibility of state function renormalization. That is, the renormalization is possible due to the fact that the evolution of spontaneous infinitesimal and finite-size perturbations have the hysteresis loop, or in other words occurs on different trajectories. The presence of a relaxation time emerging as a result of finite-size disturbances leads to the conclusion that the closer the system is to the point sol-gel type transition, the more of the phase space is filled with chaotically fleeting (temporary) "gel blocks," "mindful of the future" gel state. So in some sense, indeed the "future" state of the system partly determines its "present" and the convergence of the "present" state, the function of ensemble of probabilities to the "future" "selected" state. In the case of polymers or polymerization, the selected state is a state of supramolecular structures emergence, such as a gel or any other that has one common quality, viscoelasticity.

The renormalization process can also be described in terms of nonlinear analysis. For nonlinear systems at the branching points on the evolutionary trajectories, that is, bifurcation, points are important in the scale of fluctuations (perturbations) of the system. Depending on which of the many possible paths are taken, the system will move from one state to another. For some nonlinear (in particular viscoelastic) systems with the same values of the parameters, infinitesimal fluctuations dissipate, and the fluctuations of finite size (large enough) transfer the system into another new state. Under certain threshold conditions due to random or nonrandom external finite perturbations, the system can select a *special* in relation to infinitesimal perturbations the way of development. In this case, depending on the trajectory of the system and the scale of perturbations, the system can move from one state to another at different values of parameters. In this case, one speaks of hysteresis trajectories between bifurcation points.

Near the bifurcation points in the system, they may form many dynamical microstructures, as "building blocks" of the future states, fractals. From most of these prototypes of future systems, fractal structures, survive those microstructures, "blocks" of the future states, which are the most stable under the given environmental conditions. In general, the whole process is random and uncertain. However, in the hysteresis loop between bifurcation points the system is in the so-called deterministic chaos. The most stable (surviving in the competition) from fractal structures becomes a macrostructure called a strange attractor. The formation of a strange attractor indicates the system is in a qualitatively new state. In this case, one speaks that a system is in a state of deterministic chaos.

The system in the new state can be divided into many strange sub-attractors due to branching, e.g., chemical reactions. And state of every new sub-attractor is in the area of hysteresis bifurcations loop. Thus, each strange sub-attractor is a segment of the evolutionary path from one bifurcation point to another in the hysteresis loop. Figuratively speaking, in our situation a strange attractor can be compared to a cone, which faces its widest part to one singularity, the point of bifurcation, and the vertex to the other, i.e., to a new state. If the system falls to the hysteresis loop of bifurcations of the strange attractor of the new state, it evolves to it. All the trajectories of the system in phase space shrink to a single point at the apex of the cone. In each previous state in the general case, the system is relatively more chaotic in relation to the next. As a measure of the degree of chaos here, I use the definition of this concept given by Yu. L. Klimontovich [29]. One of the central points in the work is the definition of the degree of chaos (order). In this work the degree of chaos is defined in relation to the type of self-organization processes. Such state is more chaotic, which is farther in phase space from the state of "self-organization." Thus, the concept of "self-organization" is proposed as the main one. The work defines the universal quantitative criterion to compare the two states according to their degree of chaos.

Also, according to Klimontovich, the Boltzmann-Gibbs-Shannon (BGS) entropy, in addition to its traditional role in the statistical and information theories, can serve as a measure of the relative degree of order in the processes of self-organization. BGS entropy is a measure of diversity necessary for the existence of natural selection in the process of biological evolution. Evolution is based on self-replication errors (mutations) in a viscoelastic medium. The presence of these two properties—a nonlinear process in a nonlinear medium—is enough to cause a molecular matrix system with progression of complexity, i.e., with an increase of the relaxation time of the medium. Different levels of the hierarchy in biological systems assume different levels of self-organization

In addition, referring to the increasing complexity, it is necessary to mention the findings of one of the founders of cybernetics, John von Neumann, in the theory of self-reproducing automata. It turned out that the capacity for self-replication depends on the complexity of their organization. At the low level, the complexity is degenerate, i.e., each automata is capable to reproduce only a less complex replica. There is, however, a quite certain critical level of complexity at which this tendency to degeneration is no longer universal. To living systems, such critical level is the threshold of complexity that is responsible for the emergence of the simplest viscoelastic matrix among the reacting molecules The complexity of the structures beyond this level increases..

I have to note that any small fluctuation near the stable stationary state will manifest itself in the "excess entropy production." However, large fluctuations—mutations that alter the system itself—may have a different sign of the change in entropy. The emergence of mutants corresponds to fluctuations of entropy production, the sign of which is determined for each mutant by its influence on the increase or decrease of the relaxation time of the system (at constant kinetics). With the growth of the relaxation time through mutation, the fluctuation of entropy is negative. According Glansdorff, Prigogine [84], such negative fluctuations should first lead to a total or partial collapse (chaotization) of structures of the existing stationary state and ensure transition to the new stationary state, or, in other words, to ensure the evolution to the new higher level of the organization with increasing regularity. Increased internal order is an increase in the share of organized polymer sequences or supramolecular structures in the system, i.e., the degree of complexity (relaxation time). If the mutation does not lead to the complication of the structure, the fluctuations of the entropy production is positive.

However, the situation is not that simple if we assume that the different structures or polymer sequences have different levels of free energy and, therefore, have different reaction kinetics of the structures' formation and disintegration. In this case, it may change the balance of entropy in this fluctuation. Yet a crucial parameter for the selection of mutations remains the same as the maximum life span of the state or, in other words, the magnitude of the negative entropy production. Theoretically, it is possible—for example, such a change of kinetics of the processes with a decrease of entropy production—that it can be more than covered by the growth of negative entropy production at lower relaxation time. But in any case, this relaxation time of the system in the new evolutionary state will be maximal for a given kinetics of processes.

Here it is appropriate to remind readers that among the fundamental interactions of nature (strong, weak, electromagnetic, and gravitational forces) there are no chemical interactions. Chemical interaction is a short-range component of the electromagnetic interaction that stands out in gases, liquids, and solids by intermolecular interaction. Naturally, the chemical interactions exist only for molecules to contact each other. In the cells of a living organism, molecules are involved in numerous biochemical reactions, providing its vital functions. The only thing that cannot provide chemical interactions is to keep the processes correlated on the scale that exceeds the diffusion path length of molecules in living tissue. To implement these correlations across the body at each hierarchical level of its structure requires symbiotic chemical and mechanical interactions having a relatively large long-range in time and space. For each hierarchical level of the organism, such mechanical property is viscoelasticity, which is determined by the characteristic time of mechanical relaxation in the system. Each hierarchical level of the living body structure has its own characteristic relaxation time (time spectrum). The presence of the relaxation time in the system provides to the system a memory of the system state for a relatively long time (much longer than by Newtonian fluids). This memory provides to the system long-range processes interactions by transferring the memory of the system states in a relatively distant past at a relatively large distance.

Symbiotic mechanochemical forces are in organisms based on auto-oscillatory reactions, like BZ reaction. Mechanochemical properties are different for different levels of the hierarchy in biological objects. Hysteresis of bifurcations in the living biopolymer viscoelastic systems integrates interaction processes at all levels of the hierarchy. Processes at higher levels of the hierarchy in biological systems control processes at the previous level.

It is also important to note that the hysteresis loop of trajectories peculiar to viscoelastic systems is not only near to bifurcations. For example, loading and relaxation of viscoelastic fluids and solids happen in the different (hysteretic) paths. So the physical state of an ensemble of particles in dynamic (reacting) systems in the gel formation bifurcation hysteresis loop includes the simultaneous effects of two "pasts." The "first past" is a "chaotic past." The "second past" consists partially of the ensemble of probabilities of microstates leading to the "future" macrostate. Thus, the "second past" of the system partially filled with its "future" affects its present from the maximum depth of the past equal to the relaxation time of the "future state."

In biological systems, the influence of the "future" is manifested in the presence of so-called "goal-setting" in the system. I would like to note that the terms "future" or "goal-setting" are used here without any mystical meaning. These terms do not have anything to do with any physical movement in the time or "cognitive coherence" of molecules, attributed by some researchers to molecules in the evolving system. Taking into consideration the viscoelastic hysteresis loop in living systems may explain why any substance that enters the body through its metabolism is at certain stages of chemical transformations "channeled," statistically speaking strictly to certain parts of the body to perform its specific functions. This is one of the mysteries of nature called creodicity, which lies behind such things as the origin of life and the functioning of all organisms. In biology, fifty years ago, the term creod was introduced by D. S. Waddington, a fact that is half-forgotten now. According to Waddington [13], "creod" is the trajectory of developmental change (arising). Creodicity, which can be defined as "channeling" for branching paths of the cascade of chemical reactions in all the structures of living cells, is a result of the existence of viscoelastic hysteretic interactions between all the hierarchical structures. The same is related to the whole body and to any of its parts. Dynamic self-consistency and internal coherence of motion in biological systems surprise researchers. Self-consistency, in this case, means that from all the possible paths of chemical reactions in the phase space of the metabolism of living organisms, only those that are self-consistent globally in this space can exist. This is defined by the existence of a strange attractor of metabolism. Cascades of self-consistent trajectory of metabolic reactions, I call creods. Creods form in the space of cells and their remarkable coherence is explained by a single physical phenomenon—hysteresis of bifurcations of autowave processes in viscoelastic biopolymer matrices of each cell structure.

On each level of the hierarchy of living systems, its "present" is a "future" for the ensemble of states at the previous level. Due to the fact that all the states of particles of the previous hierarchical level are in hysteresis loop of bifurcations of the "present" state, viscoelastic renormalization determines the ensemble of probabilities of particle states leading to this present state. The set of branching sequences determined the states of ensembles of particles in these hysteresis areas forming a metabolism of the body or its parts. It should be noted as well that this is not only about the transitions to the gel state. In the phase space of the metabolism of the body may emerge spatial-temporal structures, which exist because the viscoelastic properties of the medium can influence the chemical and diffusional processes on the previous level of the hierarchy. That is confirmed by the already quoted results of work [7, 8] for the hysteresis type of bifurcation in the process of heat and mass transfer in non-Newtonian liquids. The existence of such a very special type of this dynamic determination of the states of parts of an organism by the state of the whole organism is a property inherent mainly (if not only) to living systems.

The necessity for a process of renormalization existence in living systems can also be defined from the different approaches to the problem. For example, the theory of general dynamical systems based on the topological duality principle also describes the dependence of the "present" of the system on its "future" state [14]. As a topological invariant, it is considered the phase space, with the exception of its part occupied by its "future" state. This parameter is a kind of reverse parameter dedicated by "future" state function. It seems to me there is a deep inner connection between the different descriptions of renormalization. Moreover, I think it's very meaningful to use the topological methods for describing creodicity phenomena.

It should be noted also, as demonstrated in a different approach in other work [15], that the evolution of physical non-Markovian systems with fading memory—i.e., having the relaxation times—is described by equations with fractional derivatives depicting some kind of fractal structure with channels. Channels structure—analogue to renormalization—may be different and may be generated by a specific fractal structure of the medium. Such processes are classified as processes with the "remaining" memory. In our case, a channel could refer to a different rate of diffusion through a gel formed of molecules of a different "size" and, possibly, to a description of creods in a phase state of the metabolism of living beings.

On the other hand, in statistical physics, one of the criteria of the irreversibility of the process is the change of sign of the time. As stated in work [15] in the specific process described in the fractional derivatives, with this change, some states of the particles are preserved, but others correspond to an irreversible loss. According to this work, in the relatively slow relaxation process (relaxation time is substantially greater than the characteristic time of the process), when the magnitude of stress changes more slowly than its first derivative does, irreversible loss of some of the system states occurs. Loss of the states in such a description naturally takes into account the irreversibility of nonlinear processes. This is somewhat analogous to a renormalization of ensemble probabilities of possible states of particles in phase space with the hysteretic type of direct and inverse transitions (bifurcation) observed in nonlinear viscoelastic media in the processes of heat transfer [7, 8]. Any state of a system consisting of an ensemble of particles is described by the so-called "wave function." The square of modulus of a function of the state for any particle is a probability density of its existence in this state. The implementation of such a special state of the system as a gel can only be by a very narrow wave function of state. In order to carry out such an unlikely form, the functions of the state of individual particles should be renormalized. That is, as closer parameters of the system approach to the state function of the gel, more and more particles (building blocks), which may lead to the formation of the gel, take the values of parameters corresponding to the gel phase. They get this gel building blocks state relatively quickly, but they lose that state relatively slowly in the viscoelastic relaxation.

When the appearance of new "blocks" occurs frequently enough and their relaxation time is large enough, the whole system goes into the gel phase. Or, in other words, molecules reform supramolecular structures. A necessary condition for the formation of these structures on the proposed scenario is the existence of hysteresis of their emergence and decay. The wave functions of the particle states in the hysteresis region as it approaches the gel point are seeking the identity with a "future" state function of the gel phase. At identity, in the hysteresis region, the degeneration of the state function happens for some parameters. These parameters are the physicochemical parameters of gelation.

A structured "future," any other condition being equal, is solely dependent on external boundary conditions. Indeed, if we look at an example of such a system in the form of a gel arisen from a simple monomer, we can conclude that the conditions of gel formation and collapse (bifurcation point hysteresis) are determined by the physicochemical properties of the monomer, solvent, and boundary conditions. The variation of the boundary conditions can theoretically lead to variations in forms (for example, the degree of polymerization between nodes of the gel matrix) and bifurcation points of the gel.

In the hysteresis points, the system will make two types of transition: chaos to gel and gel to chaos. A characteristic feature of the transitions from sol to gel and the collapse to a new state of the gel is its extremely high speed. That is, gels in these points are extremely sensitive (exponentially) to changes in the ratio of solvent-monomer, temperature, and possibly other parameters. If any binding agents enter into such gels (e.g., ions) that change the mechanical properties of the gel matrix, a situation may occur where the system experiences a so-called auto-oscillating reaction. These reactions have been observed experimentally in the form of gel layers mechanically pulsating for a few hours [76-78]. If, after the exhaustion of the chemical energy in the gel, a mechanical kick is rendered, then it again is excited and goes for quite a long period of oscillations. This property of polymer gels is extremely important to the process of evolution. The emergence of auto-oscillating mechanical pulsations (oscillations) in the primary gel under the influence of external ion flow and the variation of this flow will change the characteristics of pulses (e.g., frequency) and diffusion, binding, and unbinding (replication) of oligomers. Furthermore, the imperfections of replication processes appear as mutant replication. The combination of the above processes and mutations should lead to the formation of new types of supramolecular structures and gels. That is, there is the beginning of the evolution of supramolecular structures determined by physicochemical properties of the system at the previous level of the hierarchy and by mutations and boundary conditions. The speed and direction of the evolution of supramolecular polymer structures are extremely sensitive to the boundary conditions, That is, for example, to the position of a small domain of the gel, in the total volume occupied by the gel. If the gel is formed of a certain thickness on the surface of the water, then along with its thickness will be the changed water pressure. Or on the surface will be different mechanochemical boundary conditions that lead to the different conditions of molecular and supramolecular evolution. Or a lot of other environmental factors may be involved, the effects of which are the subject of special studies far beyond the scope of this book.

With the ongoing processes will arise complication of the gel structure. It will become heterogeneous. With stirring, under the influence of external factors due to the hysteresis type of transition path sol-gel-sol, it is very likely to maintain various gel structures in the mixture. This rapid complication of conditions for replication, diffusion, reaction conditions, molecular, and steric factors in heterogeneous matrices will accelerate the evolution through the production of a wide variety of new molecular species.

Changes in the external environment and the diversity of molecular building blocks for gels can lead to reformatting their matrices. This in turn will lead to the emergence of new types of molecules and supramolecular structures of many types, not necessarily only gel type. For example, there may be liquid crystalline structures, etc. Ordered crystal structures could theoretically serve as templates for the assembly of new molecules that provide the necessary compartmentalization. Due to viscoelastic phase separation (spinodal decomposition) [20], inherent to the polymer solution, inside the crystals there will be changes in the conditions of diffusion, condensation on the surface, the concentration of monomers, polymers, and solvent - water.

In order for such a system to retain the ability for continuous evolution, i.e., successive passage through novel structural states, it is necessary for it to stay in the hysteresis region of the next state. Perhaps this condition corresponds to the existence of an Eigen limit cycles in living systems. This gives rise to a sequence of hierarchical structures that store up ("freeze") the negative entropy of the system. As soon as they fall out of the sequence chain, the system immediately begins to degrade structurally and cannot exist as a single unit. For example, one can imagine that bacteria, which are systems with some evolutionary path, include many of the hierarchical transitions, resulting in gained ability to mitosis. At this point, life has moved from physicochemical on to the population type of evolution. Population evolution is essentially molecular evolution occurring within cells. But the evolution of the population is not concerned about the survival of the individual but a population. At the same time, we have to consider that evolution is not driven by what is good for species or the population, but for tiny organelle, called nucleus. Or, to be more precise, for the macromolecule called DNA.

Formation of multicellular organisms combined population and individual evolution. In multicellular organisms, there are many types of cells with different functional roles. Having the same genome, they are still epigenetically different. Epigenetic evolution during the life of an organism is called its development, and adaptation ensures the survival of the organism. Genetic mutations in the DNA in the cells of reproduction ensure the survival of the population.

10
Viscoelastic hypothesis

For the reasoning of the high probability of the viscoelasticity emergence in a primordial soup, I have suggested my own hypothesis for the emergence of life. According to this theory, when life was born out of the primordial soup, the simplest self-exciting oscillation reactions led to the formation of matrices from the first oligomers (relatively long molecules). This most likely happened in an isolated volume with the participation of amino acids, nucleotides, lipids, and other substances. Oligomer molecules spontaneously and irregularly formed a primitive matrix structure of the primordial gel. When a new long molecule is formed based on the existing molecule, there is a local decrease in entropy in the solution. Also, because of the so-called dynamic asymmetry of diffusion of the solvent and the polymer, viscoelastic separation of phases occurs as well. This viscoelastic phase separation helps to convert the negative entropy associated with energy flow and composition of the system into negative entropy associated with configuration entropy of macromolecules.

It should also to be noted that the basis of a viscoelastic gel could be not only peptides based on amino acids, but on lipid molecules as well. As shown by many researchers, the abiotic synthesis of lipids is possible in primordial conditions. Lipid bilayers have viscoelastic properties and can exist in both phases as liquid and as gel. Additionally, the lipid layers typically have a porous structure with the existence of channels for molecular transport. Theoretically, the formation of lipid gels can provide the appearance of viscoelastic properties of the medium required in the concept of life's origin I have advanced. However, this does not change anything practically in the subsequent arguments. Moreover, lipid molecules and oligomers might create intersecting matrices or lipid molecules may cover filaments of oligomer matrices and establish certain conditions for the reactions and mass exchange in the regions of the viscoelastic phase separation required for the evolution of molecules and structures.

As well, authors of one study [16] proved that not only lipids but peptides also can organize themselves as bilayers, and they have generated the first real-time imaging of the self-assembly process. But there are some other possibilities as well. Duplicating the harsh conditions of cold interstellar space in their laboratory, NASA scientists have created primitive membranous structures found in all living things [17]. These chemical compounds may have played a part in the origin of life. In the lab, the scientists recreated the conditions found in space—which is a cold vacuum—by zapping a series of simple ices with the ultraviolet radiation found everywhere. They created solid materials that, when immersed in water, spontaneously created soap-bubble-like membranous structures that contained both an "inside" and an "outside" layer. Those bubbles, theoretically, may have viscoelastic properties for beginning the process as was described above.

It seems like the replication of molecules happens chaotically in the beginning. That is, almost every newly formed oligomer molecule from peptides or nucleotides can be considered a mutation. It is because random replication is forming different oligomers each time. The more stable mutations are inherited, are repeated multiple times, and create new mutations. This is the beginning of the biological evolution and self-assembly of the primordial gel, which is generally observed for such systems—for example, in the processes of polymerization. However, self-organization can occur in nucleotide oligomers by diffusion through a highly elastic peptide gel matrix.

The appearance of following mutations is a branching process, and the influence of preceding mutations on this has a characteristic similar to the flicker-noise. Due to this, a macrostructure of the system that is created is the strange attractor. However, the structure of molecular replication must be stationary for the state of the strange attractor to exist. In the primary state, the feedback mechanisms have a linear or a "mildly" nonlinear character, but because of this they do not provide enough stability for emerging structures. Examples of such "mildly" nonlinear structures are Bernard convection cells in a thin layer of liquid losing its structure immediately after the heating is cut.

The existence of such a system in the form of the strange attractor is initially provided by the viscoelastic properties of the medium (the gel), which provide strong nonlinear feedback mechanisms. Nonlinear feedback actually determines the renormalization of ensemble of particle states in the structural transition. To put it another way, the emerging macromolecular structures have a certain range of relaxation times of their mechanical states that stabilizes the system at large fluctuations of external conditions. The mechanical viscoelastic properties act as a special feedback mechanism by regulating the mechanical stabilization of the biphasic gel and of molecular replication through the asymmetry diffusion processes. To a certain extent, the viscoelastic phase of such a system is like a "reservoir" of saved negative configurational entropy. That is, in unfavorable conditions of the external environment, the processes and structures function more or less normally for a certain period of time due to existence of the relaxation time of this negative configurational entropy of macromolecules.

If we imagine the formation of oligomers from the nucleotide bases or amino acids in a viscoelastic interaction with, for example, a highly flexible array of RNA molecules or peptide gel, then oligomers may form very stable configurations of molecules. This may be because of the viscoelastic asymmetry of diffusion in the gel "pores" of molecules of different "sizes" and "shapes" change their orientation in the flow and thus lead to changes in their chemical interaction with the molecular environment. Configurational entropy has also the meaning of "informational" entropy, which is responsible for encoding in the system. The emergence of encoding is the essential difference between living and nonliving systems. The existence of negative configurational entropy in such systems narrows the ensemble of probability of particle states in phase space; that is, it ensures the flow of physicochemical transformations in the most effective way. The effectiveness of the process is determined by the fact that most of the reactions occur in a fairly narrow spectrum. Although the system remains chaotic, exclusion of part of the reaction spectrum in the system reduces energy dissipation and hence increases the energy conversion efficiency. The accumulation of negative configurational entropy in the system contributes to the emergence and evolution of fractal hierarchical structures. The changes (evolution) on the higher levels of the hierarchy affect (in different degrees) the ensembles of probabilities of states of the particles on all the lower rungs of the hierarchy. This in turn provides a unified evolution of the system as a whole at all levels of the fractal hierarchy. Viscoelastic feedback loops in nanoscale, microscale, and the scale of the whole system (different relaxation times) are the main contributing factor to the functioning and evolution of such systems.

We have to notice as well that the energy needed to create macromolecules comes from metabolism. It means that there exist some energy-consuming trajectories of metabolism in which macromolecules of polymers are made. The source for these first macromolecules might be amino acids peptides or nucleic acids that presumably existed in the primordial soup. Most of the macromolecules are fairly unstable and are subject to destruction. But because, after creation, macromolecules immediately arrange themselves into some sort of (probably very primitive) supramolecular structure, their chemical destruction takes a bit more time than synthesis. Thus the concentration of macromolecules in the reaction's domain is maintained by a balance of synthesis and destruction, that is, by a balance of the energy fluxes that pass into the system and out and synthesis/destruction rate.

11
Nonlinear evolution of life

The basis of evolution is mutations and competition. In such relatively simple systems, these happen on the molecular level. The rate of replication, as in any evolutionary process, provides a larger number of individual molecules and improves survival. The amount of substances and energy is regulated by the laws of conservation. However, the exchange of energy and matter in live systems is also determined by the second law of thermodynamics, according to which in the process of exchange there is a certain irreversible loss of free energy as a result of increasing entropy. Emerging competition leads to the transformation of negative entropy associated with the flow of mass and energy to negative configurational entropy of the macromolecules' structure. We can assume that in the early processes of metabolism, the main role was not played by the existence of some long sequences of reactions, but more likely by the branching of trajectories in the form of a high frequency of mutations when molecules replicate. This mechanism provided a very fast evolution in the primitive stage because the stability of such a system was determined, at least partially, by a continuous changeability. In other words, the first stable live systems were very different from the modern structure of a stable system of metabolic reactions. They existed as a constantly evolving structure:

the necessary sources of bifurcations were mainly mutations and the nonlinear feedback mechanisms were provided by the viscoelastic properties of systems. It seems that the evolution of viscoelasticity during the gelation reactions consisted of modifications of the hydrophilic/hydrophobic balance. This allowed the foundation of hydrophobic interactions and hydrogen bonding. So why is nonlinear feedback so critical? The thing is, just the existence of long molecules in such a system does not fully differentiate live matter from nonliving matter. What is crucial is the evolution of molecules and the hierarchical structures that are based on them—and that only appears when there is nonlinear feedback. It looks like the relaxation time of these structures can be subject to evolutionary selection. The growth of the relaxation time is observed in polymer systems with the elongation of the macromolecules and/ or their branches. However, we must recognize that the natural restriction for the relaxation time growth is kinetic effects. So with the evolutionary development of state of the system with the maximum relaxation time is the glassy state. But in reality, this condition will prevent the reaction and the evolution stops. Only the proper balance between the reaction kinetics and the relaxation time of the system allows it to evolve. The mechanochemical nonlinearity of the medium determines, for example, the existence of the so-called nonlinear signal enhancer and various forms of resonance. In such a mix, more complicated structures can evolve.

It seems that the "live" medium of replicating and interacting molecules was in itself quite an extensive structure of a three-dimensional gel, similar to mucous in modern organisms. Through the mesh of the gel matrix, thanks to low viscosity on the nanoscale, there was a free delivery of new low-molecular elements (e.g., amino acids, nucleotides). Based on the chemical or physical interaction with the molecules of the matrix, these elements built new oligomers on the gel structure, some of which were integral to its configuration. At the same time, there could well be the emergence of coding oligomers of nucleotides, the first RNA. This process is very similar to the normal process of gel formation during the polymerization of monomers. Such systems belong to the class of systems in which, along with the replication reactions and diffusion, there are intermolecular Van der Waals forces at work. A distinctive specialty of such systems is that the structures emerging within them can be smaller than the characteristic length of diffusion (the typical distance that can be covered by a molecule from the point of its birth to the point where it will take part in some reaction). That is, microscopically and nanoscopically sized structures that are characteristic of biological systems [18] may materialize.

As a matter of fact, the phenomenon of microscopic self-organization and ordering in the systems of interacting molecules is a well-known fact in the modern physical chemistry of polymers. Periodic structures in such a system could emerge in many ways. For example, we can consider a macromolecule that is composed of two or more chemically incompatible monomers, chemically bound at their ends. Such macromolecules are called copolymers. Chromosomes are an example of a copolymer, exhibiting rigid regions alternating with semi-flexible regions [19]. In a solution or a gel of such copolymers, we would expect a phenomenon: a viscoelastic phase separation in polymers [20], but occurring locally and producing complex periodic structures. Alternatively, in other research [21] it is found that in the initial state of viscoelastic phase separation, an increase in the relaxation time tends to increase the rate of growth of the structure factor, and tends to decrease the wave number of the peak in the structure factor. This means that the sizes of spontaneously emerging structures of gel during its evolution—i.e., with increasing elasticity—decrease but appear faster.

12
The birth of Gaia

It seems as if early life was in the form of layers of constantly evolving viscoelastic gel on the bottom or surface of the oceans or on the surface of "fresh water" ponds, or in porous volcanic rocks in water. Whether it was the bottom or the surface depended on the source of energy, either the sun or the thermal springs from the mantle. The source of chemical elements is also important: this could be the waters of the primordial ocean, or the highly concentrated mineral springs, or fresh water minerals dissolved in porous rocks. Moreover, the morphological and rheological behavior of the self-assembled molecular network formed in this gel is presumably related to the thermodynamic character of the gelation in given external conditions. We can also view such systems in terms of cybernetics. A fundamental phenomenon that is common to polymer molecules is the ability to build new signaling molecules. The links of the gel matrix are a network of nonlinearly connected oscillators with the capability to transfer signals. Modulation of the signals by the properties of the system (iteration) is a change in the frequency of mutation. That is, the polymer matrix is a system which records, transforms, and translates information. But it is a quality of some cybernetic systems in which primitive structures have appeared, thanks to molecular evolution in which there are constant mutations. Such a gel or matrix can be viewed as a lot of very simple cybernetic elements, where each one is capable of performing a small number of operations and keeping data about a few proceedings. In mathematics such structures are called cellular automata, which carry out analogue processing of information. Analogue devices are not programmable in the general meaning of this word. The programming of their function is determined by the physical-chemical processes that take place in them. The preferential assembly of some types of molecules by replication was happening in these analogue structures, which corresponds to the appearance of the primitive genetic code. Such open self-organizing systems, in cybernetics, belong to the class of adaptive systems. In the process of evolution, they have the ability to adapt to changes in their internal conditions. They can continuously alter their structures and the algorithms of their function as well. Besides, this information processing on a molecular level is a well-known fact in biochemical systems. As an example of similar molecular computing, we can consider the process of the regulation of messenger RNA (mRNA) synthesis by special proteins called transcription factors. This makes it possible to theoretically implement an arbitrary logic relation between the transcription factors and the result of synthesis. Implementation of logic actually establishes a platform for in vivo molecular computing.

The probability of the normal evolutionary emergence of, say, the relatively long RNA molecule, as was mentioned by many researchers of the subject, is close to zero. But thanks to "cybernetic evolution," the process was going quickly relative to any other scenario due to the "explosive" mechanism. It is possible that the results of one study [22] are a distant illustration of such a type of fast evolution on the molecular level. In that study researchers developed a special method to artificially induce accelerated evolution of enzymes in a test tube, which enabled them to engineer "tailor-made" enzymes. The method is based on introducing many mutations to an enzyme, and the mutated versions were scanned to select for those exhibiting improved efficiency. These improved enzymes then repeatedly underwent further rounds of mutation and selection for better efficiency. This method can make enzymes more productive by factors of hundreds and even thousands.

In order for the cybernetic analogue system to function adequately and quickly, there needs to be a certain directionality of the processes. Otherwise, the function of such a "computer" will boil down to the banal processing of variations that even at top speed can take too long compared with the time of the existence of the visible universe. In the study [22] mentioned above, the selection was carried out artificially. However, a different study [9] reveals that for nonlinear, nonequilibrium systems with a limited cycle (the iteration processes converge) and a hysteretic character of bifurcations (viscoelasticity), we can determine the sequence of microstates (process of "calculation") by the "future" macrostate of the system. When forming long molecules (oligomers), they can create matrices with physical or weak chemical bonds.

The formation of the first matrices means the formation of the first structures. The conditions of existence and interactions in such a matrix will be different from the initial conditions when it did not yet exist. The matrix will only accept molecules that, in one way or another, conform to the conditions of their stay in the matrix. Other molecules either stay in the medium, slowly saturating it, or leave the matrix. This is the emergence of a simple feedback mechanism from the rejected molecules. In turn, this changes the primary process of oligomer formation itself due to the altered chemical conditions. This phenomenon is mathematically called the renormalization of probabilities of ensembles of microstates of molecules by the final macrostate of the system [10]. So the method of renormalization of polymeric gels could explain the appearance of periodic structures, as a study [23] conducted by H.C. Öttinger and his colleagues confirms. In recent theoretical proposals for the process of critical dynamics at polymer gelation [23], H.C. Öttinger suggested a dynamical renormalization approach within a general framework of nonequilibrium thermodynamics. By this method, he could demonstrate that the self-similarity of the structure close to gelation can be directly connected to a critically high relaxation time of viscoelastic gel arising.

A good analogy is the autocatalytic reactions, when the result of the reaction depends on its products. And this is not only an analogy; there are many reasons to assume that the reactions of the described type of molecular replication happen along the autocatalytic pathway [24]. This is what gives our Gaia "computer" such fast processing speed and the renormalization of probabilities that are mentioned in [9]. In the autocatalytic reactions, the set of possible ensembles from molecular replication is quite narrow due to the nature of the process itself. Moreover, in autocatalytic processes the reactions distribute themselves at a much greater speed than in other, similar systems. So, according to [24], for a general chemical process with a reaction rate of 1×10^{-5} s/molecule, it would require around twenty billion years to generate a mole of reaction product. In contrast, it only takes 79 µs in an autocatalytic process to produce the same amount of product. But even so, autocatalytic processes alone are not enough to support the conditions for the emergence of life. The viscoelastic feedback is also a key factor.

But what chemicals in the primordial soup can catalyze the autocatalytic process of replication? The possible answer is some small organic molecules. The smallest RNA enzyme—ribozyme, a form of RNA with only five nucleotides that can catalyze chemical reactions—is described in an article [25] published in the *Proceedings of the National Academy of Sciences*. According to the author of the work, "it appears that the first catalytic macromolecules could have been RNA molecules, since they are somewhat simpler, were likely to exist [or could be formed] early in the formation of the first life forms, and are capable of catalyzing chemical reactions without proteins being present." The existence of that kind of small-molecule organic catalysts could support initial autocatalytic reactions in the primordial soup, leading to the formation of gels from relatively short oligomer proteins or RNA molecules according to the viscoelastic scenario of prelife and life origin presented above.

13

Viscoelastic autocatalysis and self-assembling

Let us try to explain the meaning of renormalization by using an example not far off the described physicochemical system. Suppose that in the primordial soup there is an area where various chemical reagents are delivered. These reagents can take part in reactions between themselves and/or with products of reactions. We should also assume that these reactions result in the spontaneous formation of oligomer (relatively long) molecules. During the course of the reaction, oligomers accumulate where the reactions take place. Physical and chemical bonds form randomly between them, and when a certain concentration is reached, a primitive matrix develops.

Imagine as well that some oligomers, due to their special properties (chemical composition, etc.), can leave the reaction zone through the cellular mesh of the matrix (by diffusion, for example). Let us also suppose that there are molecules that cannot leave the reaction zone. As the concentration of such molecules increases, the reactions that create them are inhibited. At the same time, the concentration of molecules that can leave does not inhibit their production, as their concentration always stays low. In this case the set of ensemble of the possible states of the system (participating reagents, types of reactions, and the types of oligomers being formed) is significantly smaller than the set of ensembles of the possible microstates of the system at the primary stage. This is because relatively few variations of oligomers can be created, and the diversity of chemical reagents and reactions these molecules could be involved in is also reduced. It does mean that in such systems the emerging molecular organization is not only a direct consequence of the reactions pathways involved in the assembly process, but also that reactions pathways are changing from the influence of viscoelastic feedback of evolving supramolecular structure. Thus we can say that there is a renormalization of probabilities of the ensembles of the microstates existing by the macrostate of the system. In terms of nonlinear analysis, this can be called the first iteration. I want to note that I have not made any improbable assumptions and that this is quite possible to have been happening in the early evolution of life on Earth. If there is some sort of autocatalysis, which can appear for many reasons, then the speeds of reactions in such a system can increase considerably and narrow the spectrum of ensembles of microstates.

In the hysteresis loop after transition through the bifurcation point the macrostate of the systems in the new phase controls the state of the ensemble of microstates in the previous phase. But this situation is very similar to autocatalysis, when the result of the reaction has a reinforcing effect on the reaction itself. That is, because of renormalization, the autocatalytic state of the system occurs without a catalyst. The narrowing of the function probability of states distribution in such "viscoelastic" nonlinear autocatalysis is significantly reduced, for example, the number of variants of sorting for simple operational RNA. So, apparently, removed is one of the problems of the origin of life, which is that the random emergence RNA could not happen during the lifetime of the visible universe.

I think that the difference between that form oligomer molecules, or RNA, or the protein that became an arbitrary genetic code is not in the specificity of its energy states, but in the fact that the balance of persistence and variability of its chemical and spatial structure in the given physicochemical conditions is consistent with the capability of existence of a multitude of closed-cycle reactions required for self-replication and mutation. And of course also the existence of viscoelasticity of the environment is necessary for evolutionary selection.

There is no mystical code of life that is given to us from somewhere "above." There is only the external and internal conditions that are imposed on the possible chemical reactions and physical processes of products transportation for a particular class of molecules and structures, for example, primitive nucleic acids. As a result of chance, but with a considerable probability through viscoelastic renormalization in the system, there is chosen a narrow pool of structures that are capable of providing themselves the products of molecular replication and their transportation with possible interactions in the pool.

Reproduction-replicating molecules create "ecological" issues, and with their help destroy intermediate molecules (limits their reproduction) and at the same time due to imperfections of replication processes are creating new types (mutations). New types of molecules arise, not so much as the result of any special optimal criterion, but rather as the mutations that have survived. The very fact of survival is the criterion of an emergence of the distinctive novelty in the system. It is as well the evidence of the existence of a suitable ecological niche. Such a complex set of parallel processes presumably can lead to the synthesis of something similar to the genetic code. The existence of the code provides a relatively stable product of self-replication. Crucial in this is the presence of a viscoelastic feedback and the renormalization process as a selective factor necessary for the start of Darwinian evolution of molecules.

It is also important to note that the initial gel-matrix formed by oligomers will perform the function of a parametrical filter because the diffusion processes within the matrix are dynamically asymmetric. That parametric filter regulates diffusion in the solvent phase depending on the effective size of the molecules. This means that at the exit point of such a reaction area we have a modulated distribution of molecular concentrations according to their physical and chemical properties. This leads to the next iteration: the formation of more complex structures. For example, it could lead to a denser packing of oligomers, or to matrices with more regular configurations, or more cross-linking bonds, etc. As the system becomes more complicated, the variety of ensembles of possible microstates narrows. This brings about the materialization of structures similar to those structures formed from a class of amorphous oligomer molecules known as "intrinsically disordered proteins" [26]. Unlike typical proteins in the cell, intrinsically disordered proteins do not adopt a stable globular form in isolation. Rather, they are like a messy, unfolded string of yarn, whereas typical globular proteins more closely resemble yarn neatly knit into complicated and functional shapes like that of a glove. The equilibrium states of molecules in such an ordered structure are known as an alpha helix (like the coil of a phone cord) called the F state. Researchers [26] have seen how the protein shape changes soon after binding to its partner molecules. This means that in our dynamic conditions of a continuous supply of new molecules through the initial matrix, similar new states might provide the system with many more modes of regulation and selection of incoming molecules. Each molecule aggregates differently within different structures, and thus has unique properties and modes of interaction with other long and small-molecule partners. The rise of viscoelastic properties in oligomer matrices considerably complicates the character of molecular interactions in such a system. In terms of the dynamics of nonlinear processes, there is an addition of viscoelastic nonlinearity of media in the equations describing the processes.

Also note that, the equilibrium concentration of the polymers, which could be expected in a primordial soup, according to most researchers [27] is so low that they are at a thermodynamic tendency to dissolve in water, but not to grow. This fickleness of polymers in water, as well as the chirality of the main molecules of life, is "a headache" for researchers who are working on the evolutionary ideas of the origin of life. However, the collapse of polymers in water is rather characteristic of single molecules. In the formation of the gel and/or the supramolecular structure's stability of the polymer molecules in their composition may rise substantially due to the interaction with each other in an oriented state. This tendency to decay will also be overcome by a local decrease in the concentration of water in the supramolecular structure's composition because of the existence of viscoelastic phase separation.

For the reaction necessary to the formation of linear polymer chains, there are bi-functional monomers, that is, molecules with two active end groups, which are connected with the active groups of other molecules to form a polymer. If you would have at least a small fraction of molecules with one active end group, instead of linear polymers, "branched" polymers will be formed. And the degree of polymerization will not be very high. Some researchers

[28] see it as a problem that the relatively high temperature of the primordial soup can also contribute to the formation of "branched" polymers. However, as proposed in *The Food Delusion* [1] and concepts presented here, branching and the degree of polymerization do not play a critical role. Central to the proposed approach is that the degree of polymerization should be sufficient for the emergence of a viscoelastic matrix phase. Viscoelasticity of the system will determine the emergence and evolution of macromolecules and fractal supramolecular structures due to the process of viscoelastic renormalization. Availability of the supramolecular and gel phase could theoretically lead to macromolecular evolution in any of the directions determined by external (boundary) conditions. A result of realization of a certain trajectory of evolution is possible (at least not prohibited) in the emergence of linear polymer molecules.

Due to viscoelastic phase separation [20], the stability of the matrix structure will rise. Even with the termination of the generation of new molecules, the matrix can theoretically exist indefinitely, if permitted by the external environment and the stability of the phase separation. Due to the viscoelastic phase separation in the matrix, there will be areas with different rates of diffusion of particles of different sizes. In this case, the stresses emerging in the matrix will affect the mesh size of the matrix and thus modulate the process of diffusion. Apparently, diffusion processes in such a system, thanks to the spectrum of the relaxation times of the viscoelastic gel-forming phase, should be considered as occurring in the fractal (fractional) dimension of the phase space. In such an environment, as follows from one study [29], in the simplest cases, there may be stable oscillatory processes. In a reacting environment it will look like the appearance of periodic (time-space) stable chemical reactions. The modulation of the viscoelastic properties

will affect the complexity of the structure of the gel, producing more and more complex and long aperiodic molecules involved in the processes. Aperiodic molecules are
characterized by a large number of collective spatial modes,
which are described by characteristic relaxation times for
each mode. Theoretically, spatial-temporal patterns can be
generated due to the reaction and diffusion of a number of
chemicals in Newtonian (not viscoelastic) fluids in so-called
reaction-diffusion-driven instability. The reaction kinetics
of such a system is stabilizing and diffusion, a homogeniz-
ing process. But in the case of nonlinearity of media (viscoelasticity) where the processes are taking place—because
there are hysteresis types of bifurcations—instead of ho-
mogenization of processes, there is some additional spatial-
temporal pattern emerging linked with relaxation times of
viscoelastic media.

Such systems occur at each level of the hierarchy. If the
process of reproduction and mutation has begun and there
are highly nonlinear feedback loops, the system will tend to
go to the nearest stationary state with the lowest entropy.
The level of entropy of these systems, according to [9], is
defined by the macrostate of the system. The direction to
the pursuit of a state of minimum entropy of a stationary
macrostate will be the purpose of a cybernetic system that
allows "channeling" in the iteration process, cutting into
the breeding of the molecules in the gel with too exotic
mutations (ensembles of microstates with them), which certainly will not lead to the stationary state. If this is the case,
we do not have to rely on such an improbable event as the
accidental appearance of an RNA or DNA molecules and
the four-letter code. With our approach, they appear as the
result of the cybernetic iterative process and the mutation
of complex molecules with the appearance of complex hierarchical structures with viscoelastic feedback loops. These

structures that appear as the result of this evolution change
the nonlinear dynamic properties of the system. As the system is changing, at each stage it comes to some stationary
state, a hierarchical structure, which in itself is a "building block" for creating the next structures. The appearance
of these hierarchical structures in itself is an iterative procedure during the work of such an analogue computer. It
seems that the four-letter code of the nucleotides appeared
gradually, through the appearance and evolution of two-
and three-letter codes.

Probably, at the initial stationary stage, there was
a two-letter code. The next iterations created three- and
four-letter codes. In a way this is reminiscent of the syner-
getic structuring of energy and matter flows in dissipative
processes and the appearance of physical structures. But in
this particular case, the emergence of physical regularity
has the properties of the informational cybernetic structure
in the form of the genetic code. However, we should bear
in mind that the stationary macrostate does not completely
determine the set of microstates during the channeling of
the evolution of systems. Theoretically, the "channel" can
have an infinite ensemble of these microstates or evolutionary trajectories that lead to the final stationary state. Also,
the data from a study [5] where similar systems were re-
searched shows that they have the ability to independently
generate "chaos," which is determined by the final stage
[9]. It does not depend on external noises and fluctuations,
but is determined only by the nonlinear dynamics of the
system.

Also, this approach pushes back the time of appearance of life in the form of the current four-letter code to
the appearance of the preceding two-letter code. As there
are no principal differences in the primitive life forms
that are based on the two-, three-, or four-letter codes, the

physicochemical mechanisms that take place in such a gel molecular computer need to be studied. We can only note that a big role is played by hydrophilic and hydrophobic interactions that are inherent to amino acids, as everything happens in water. In such gel structures, self-assembly is possible; this is proved by a study [30] that examined the movement of nanoparticles in narrow channels. The study shows a mechanism that leads to nanoparticles displaying self-assembling behavior in a series of experiments in vivo and in silico. The researchers found that the disturbances of the fluid induced by each flowing and rotating particle drives neighboring particles away, while the migration of particles to localized streams due to the momentum of the fluid acts to stabilize the spacing between particles at a certain distance.

In essence, the combination of repulsion and localization leads to an organized structure. The researchers also found that by simply adding short regions of expanded channel width, the particles set could be re-self-assembled into different structures in a controllable and potentially programmable way. Even more complicated structures should appear due to the deformation of channels happening according to the viscoelastic mechanism, if we view the gel as a porous-elastic medium with relaxation times of the wall. In such a system, the frequent positioning of mutant molecules and their self-organization always lead to the continuous modifications of gel properties—to the evolutionary transitions that lead to the emergence of higher-level systems hierarchy through the assembly of components. The emergence of the first small fragments that resemble the modern micro RNA and peptides complicated the structure of the gel and the structure of these fragments and peptides to the point of the appearance of first proteins and primitive RNA. As indirect confirmation of the approach described above, it could be very interesting to look at the results of a study [31] where the engineered scaffolding from RNA molecules in the cell has boosted output of enzyme biosynthesis production in two orders of magnitude. RNA scaffolds in the cell organize and concentrate the enzymes, interim products, and final products in the reaction zone exactly as described above in viscoelastic gel.

It should be noted that there is a growing body of evidence that RNA came before proteins. It is possible that RNA, like molecules, are the progenitors of the genetic code, as they exist at the expense of relatively weak chemical bonds. Therefore, their reproduction is accompanied by errors (mutations), increasing the entropy of the system. There is a new chaos. Now it is RNA chaos, at least, some of which is capable of catalytic self-renewal. It is possible that the reproducible mutant RNA is potentially suitable for cross-catalysis. The result of such a "symbiosis" is completely chaotic, because the sequences of codons in the RNA are random. The probability that in a given RNA each following link of its chain will require to catalyze the exact product of the preceding link of another RNA is relatively small. This means that replication will not exist. That is, there is another sophisticated deadlock of balance with its very considerable disorder, large entropy, calculated on the base of the number of possible combinations of codons in different RNA. However, perhaps the problem could be solved thanks to the proposed approach in *The Food Delusion* [1], which is that viscoelastic autocatalysis replication occurs in a relatively narrow (renormalized) spectrum of the mutant RNA. In this case, the probability of correct cross-catalysis or interaction with replication of mutated RNAs with each other increases. That is, memorization by the catalytic breeding goes to a relatively narrow diapason of mutants.

If the resulting mutant RNAs can best interact and replicate in some oriented position, it may form supramolecular structures. In another words, if the existence of these supramolecular structures is thermodynamically possible and it gives preference to certain mutants, these mutants by renormalization will dominate the system. Of course, it is necessary that there be the hysteresis of emergence and decay of these supramolecular structures. But despite the importance of RNAs, however, the idea of the simultaneous evolution of RNAs and peptide matrices and/or supramolecular structures in the primordial soup also cannot be completely rejected.

14
Cybernetics of living beings

It is important to note again that on the physicochemical level of such a cybernetic system the "channeling of calculations" happens because during the complication of structures and selection, not all the mutant molecules can take part in the evolution of the gel structure. Also, it seems that in complicated structures, because of structuring of chemical and physical interactions, the spectrum of created mutants must be narrowing. All of this leads to the narrowing of the set of ensembles of particle states for selection and the acceleration of evolution within each iterative "procedure."

The above approach is based on the assumption that the process of molecular evolution in the appearance of live matter is determined by the emergence of oligomer molecules capable of creating viscoelastic matrices. The structural self-organization in time and space of such processes—i.e., polymerization of monomers into polymer systems—was studied in [32]. It turns out that in such systems, as the words of John A. Pojman describe, spatial pattern formation can occur on the microscopic level to form structures reminiscent of those seen in the Belousov-Zhabotinsky reaction and biological systems. Regular spirals and bubbles can be formed in the self-propagating polymerization process. According to Öttinger publications [23], "if spatially bi-stable reaction systems are operated in size responsive chemo sensitive gels, the size changes can provide a feed-back which beyond plain reaction diffusion instabilities can be the source of new self-organizing phenomena, referred to as chemo mechanical structures." In relation to spatial pattern formation in gelation processes in [23], Öttinger made an interesting observation: "Typical phase separation leads to a two-phase disordered morphology. Multiphase polymeric materials with a variety of co-continuous structures can be prepared by controlling the kinetics of phase separation via spinodal decomposition using appropriate chemical reactions. By taking advantages of photo-crosslinking and photo isomerization of one polymer component in a binary miscible blend, researchers have been able to prepare materials, known as semi-interpenetrating polymer networks, and polymers with co-continuous structures in the micrometre range." But these are the hierarchical systems we talked about earlier, which occur thanks to the nonlinear viscoelastic properties of the macromolecular gel during polymerization both from synthetic and natural monomers—e.g., the amino acids in the primordial soup. In my opinion, Öttinger and Pojman's studies [23, 32] are an excellent illustration of my suggestion that life appeared through the emergence and evolution of the viscoelastic gel.

The equation [33] of the biphasic reactive physico-chemical system is distinguished by two characteristic times responsible for the starting of the reaction t and for the influence of the feedback reaction T. The analysis shows that when $t < T$ (i.e., when there is a relatively large relaxation time of the feedback mechanism [viscoelastic medium]), this gives rise to complicated structures in the system—oscillating dissipative structures with parameters that change in time. If we imagine that there are several relaxation times or they are represented by a continuous spectrum (which complies with the real biopolymer systems), then there is an even stronger complication of systems. And if we imagine that all this is taking place not in the biphasic model medium but in a multiphasic medium, then this guarantees even more complicated structuring of the systems that gave birth to life on this planet.

The common feature between all living beings known as viscoelasticity could be the basis for the evolution of the chemicals systems on which life is based. It could, in fact, be the essential feature controlling evolution at the prebiotic level. Because all living beings are made from the same basic chemical molecules that have more than likely been around since the beginning of life on Earth, researchers are studying the life origin chemical reactions as, for example, small sugar (glucose) binding sequences that occur in all protein and peptides like insulin [34]. The influence of emerging viscoelasticity of media could provide fundamental insights into diseases such as diabetes and cancer, and processes of the feeding. Another team of researchers from the University of York [35] studying the origin of life found that by using common left-handed amino acids, it is possible to catalyze the formation of right-handed sugars. Such sugars might evolve due to the molecular complementarity in the formose-type reaction. The chemical path known as the formose reaction, discovered by Aleksandr Butlerov in 1861, is a potential route from the simple molecules, which might have been present on Earth before life began, to the sugars essential to life. The formose reaction begins with formaldehyde, thought to be a plausible constituent of a prebiotic Earth, going through a series of chemical metamorphoses leading to many sugars, including ribose, which is a key building block in DNA and RNA. However, chemists found that in general the formose reaction produces a very tiny amount of ribose but instead a lot of other sugars that lack any biological use. This means that this reaction must take place under special conditions favorable for the production of ribose. Such conditions could theoretically exist at the lattices phase of peptide viscoelastic gel formed in the prebiotic soup. The gel lattices might have an effect on selection by chirality and atomic composition of complementary short strands of sugar molecules incoming from the reaction zone of different chemicals. Theoretically, it could explain how essential-for-life carbohydrates originated and why their right-handed form dominates in nature. For life to have evolved, we have to have had a moment when nonliving things became living. Everything up to that point, according to common beliefs, is chemistry. A main point of this chapter that is not pure chemistry, but rather mechanochemistry in viscoelastic phase, was and still is the controlling factor of the first emergence, existence, and functioning of all living beings. Researchers in the field of life origin are still a long way from being able to assemble living cells from scratch in the laboratory. According to biochemist David Deamer of the University of California, Santa Cruz [36], life began with complex systems of molecules that came together through the self-assembly of nonliving components. A useful analogy for understanding how this came about can be found in combinatorial chemistry, an approach in which thousands of experiments are carried out in parallel by robotic devices [36]. But the question arises, what worked as the "robotic devices" in nature around 3.5 billion years ago.

Another researcher, New York University chemistry professor Robert Shapiro, published a book in 1999 called *Planetary Dreams* in which he argues that the simplest kind of life may arise as a result of organic chemistry reactions and the physical processes of self-organizing systems whenever the right constituents and conditions exist: a liquid or dense gas medium (not necessarily water), a suitable energy source, and a system of matter capable of using the energy to organize itself. This case also raises the question: What is specifically "a system of matter" that is "capable of using the energy to organize itself"? Without specifying the type and properties of the "robotic devices" of D. Deamer or "a system of matter" of R. Shapiro, these approaches in my opinion do not have the heuristic value.

The approach I propose indicates that emerging visco-elasticity in a system of organic chemistry reactions with diffusion in prebiotic systems leading to the appearance and evolution of hierarchical self-organized structures is laying the foundation of life and is responsible for the work of the "robotic devices" of D. Deamer in nature, and is "a system of matter capable of using the energy to organize itself ,"as R. Shapiro suggests.

In the literature there are quite a number of objections to the mechanism of life origin in the primordial soup from simple chemicals, such as peptides, by the polymerization reaction [37]. Let us consider some of them. For example, in study [37] stated that the difficulty of activating amino acids and forming long peptides under primordial conditions is one of the great obstacles to the origin of life. And in a concentrated solution of 1 M (mol/l) of each amino acid, the equilibrium dipeptide concentration would be only 0.007 M. And first we must suppose that in primordial conditions reacting systems existed with diffusion that could greatly change dipeptide concentration due to different diffusion rates for different components of the system. Also, the formation of dipeptide gels is possible in the range of quite low concentration—less than 1 percent [38].

High temperatures on the surface of early Earth, as many researchers advocate, would accelerate the breakdown of gel. The famous pioneer of evolutionary origin-of-life experiments, Stanley Miller, points out that those oligo-mers are "too unstable to exist in a hot prebiotic environ-ment." But again, according to [28] at a concentration of 2 mg/ml, the dipeptide forms a stable aqueous gel at 60 degrees Celcius. We cannot say this is low temperature for primordial conditions.

Some researchers argue that the heat also destroys some vital amino acids and results in highly randomized poly-mers. But I am proposing in this book a model of life origin in only the very early first stage of the prebiotic system that as result of "viscoelastic evolution" may create something like very primitive proteins or RNA; only further evolu-tion of them may lead to the creation of the first life and make it suitable for polymers. The same is generally related to a problem of chirality of the amino acids.

Also, some researchers argue that amino acids in the primordial soup would be impure and grossly contaminated with other organic chemicals that would destroy peptides. Another argument is that it is a chemical impossibility for the primordial soup to accumulate large quantities of "con-densing agents" for absorbing excessive amounts of water, which could slow the process of oligomerization.

In the model proposed in this book, the emergence of peptide gel implied the existence of a viscoelastic phase separation (spinodal decomposition). In the peptide phase of gel lattices where the reactions take place, the water con-tent and the concentration of dissolved chemicals will be significantly lower than the average for a system. That is, it probably at least partially eliminates the question of the "condensing agent" and of the influence of chemicals that are hazardous for the existence of peptides and amino ac-ids. In addition, according to some researchers, vital amino acids (for example, cytosine) are too unstable to have ex-isted on a hypothetical prebiotic Earth for long, and amino acids would be too diluted in primordial soup to actually interact with each other. And even if the amino acids could have formed dipeptides, they would soon hydrolyze. The existence of the dipeptide matrix skeleton phase could help to overcome all of these problems because processes will take place in concentrated phase, forming some kind of compartment physicochemically isolated in some degree from surrounding diluted phase. Also, viscoelasticity due to the evolution of the complexity of gel structures theoretically could support the tendency to form the coded oligomers/polymers required for life as opposed to random ones. During molecular and supramolecular evolution of gel structures, the randomness of forming polymers will go down. But that is not everything! Even the existence of randomness plays an important role because the randomness is actually a source of mutations, which is the driving force of evolution in the viscoelastic gel model of life origin proposed.

15
Chirality

Since the discovery by Louis Pasteur of the mirror asymmetry of organisms (use properties of matter in one of two possible spatial configurations of molecules) as the main difference between living and nonliving matter, it has been more than a hundred years. During this period, almost any area of research (including this one) would seem to be exhausted. However, the fact that living organisms use only one of the two mirror isomers of molecules such as amino acids and sugars, and do not use the other (in the nucleic acids contained only D-isomers of sugars, and enzymes, only L-isomers of amino acids), is still an intriguing mystery [39]. The violation of mirror symmetry in living systems at the molecular level is apparently connected with the fundamental properties of living matter.

Chiral (or mirror antipodal) are objects that do not have a center and plane of inversion. Simply put, if an object is reflected in a mirror, you get an object that is not compatible in space with the original, as the left and right hands for example. Incidentally, this is the term "chirality," which was introduced to the academic community by Lord Kelvin. To chiral objects in nature belong molecules that contain a so-called asymmetric carbon atom (amino acids, sugars, etc.). These molecules have the property of chirality in the event that all four end groups of atoms (ligands) bound to the central carbon atom are different. Chiral isomers (enantiomers) of these molecules are usually called "left" and "right" isomers. They are designated by the letters L (from laevo, left) and D (from dextro, right). These mirror isomers are remarkable in that the substances formed by them have identical physical and chemical properties—the same internal energy, solubility, melting point, boiling point, etc. Their only difference is that these compounds have optical activity, i.e., they rotate the plane of polarization of the light passing through them in different directions.

In inanimate nature, left and right molecules are in equal amounts. However, as mentioned above, the molecular basis of living beings is chirally pure, i.e., we see a sharp imbalance between the left and right isomers. Thus, molecular chiral purity, genetic coding, and viscoelasticity are the key properties of living matter. Application of the basic concepts of chemical physics—the science of the kinetics of chemical reactions and chemical structure of the substance—allowed us to make the important conclusion that the emergence of chiral purity of sugars and amino acids possibly occurred under pre-cellular evolution [39]. Chiral purity looks like a relic property, which Gaia has inherited from the stages of molecular evolution of the organic medium.

What is the origin of such dissymmetry in biological material? According the one study [40], there are two main competing hypotheses. One postulates that life originated from a mixture containing 50 percent of one enantiomer and 50 percent of the other (known as a racemic mixture), and that homochirality progressively emerged during the course of evolution. The other hypothesis suggests that asymmetry leading to homochirality preceded the appearance of cellular life. This is partially supported by the detection of L excesses in certain amino acids extracted from primitive meteorites. According to this scenario, these amino acids were synthesized non-racemically in interstellar space and delivered to Earth by cometary grains and meteorites. According to this scenario, the delivery of extraterrestrial organic material containing an enantiomeric excess synthesized by an asymmetrical astrophysical process (in this case, UV-CPL radiation) is the cause of the dissymmetry of life's molecules on Earth. This material may even have formed outside the solar system. Finally, the solar nebula may have formed in regions of massive star formation. In such regions, infrared radiation circularly polarized in the same direction has been observed.

But the most acceptable scenario is that chiral purity in pre-cellular life forms in my view is the hypothesis of spontaneous symmetry breaking, which explains the emergence of self-organization properties of chirality in molecular self-organization. One reason for this self-organization of chiral molecules, in my opinion, is that a viscoelastic matrix can be formed. Perhaps there is a viscoelastic gel diffusion asymmetry [Tanaka] resulting in asymmetric synthesis, or the selection of organic molecules. This process could theoretically occur on Earth, comets, and other celestial bodies with the appropriate temperature and composition.

Also, the results of [41] are partly evidence to support this view. The authors asked the question to begin with: «Could this biological preference of a particular chirality possibly have a physical origin?» In addressing this question, researchers sought to discover how chirality occurs in the first place. Their findings confirmed that chirality can arise spontaneously, even with achiral building-blocks. Scientists used a manufacturing technique called lithography, which is the basis for making computer chips, to make millions of microscale particles in the shape of achiral triangles. Using optical microscopy, the researchers then studied very dense systems of these lithographic triangular particles. To their surprise, they discovered that the achiral triangles spontaneously arranged themselves to form two-triangle "super-structures," with each super-structure exhibiting a particular chirality. Researchers [41] have shown for the first time that chiral structures can originate from physical entropic forces acting on uniform achiral particles. They found that the minimum ingredients for specific chirality to occur are entropy and particle shape. Thus if we imagine the synthesis of amino acids or peptides in the beginning of life under the influence of an external asymmetry, it may well lead to the formation of chiral molecules.

The above-mentioned approaches could possibly explain emergence and domination of "left-handedness" in chirality for amino acids in life forms. This can occur because of the affinity of only some by the chirality of the amino acids preferable for liquid-crystal structures of supramolecular peptide gel matrix phase. Most scientists believe that the choice of chirality in life on Earth was purely accidental, possibly based on carbon alien form of life existing somewhere in the universe, and there will be another form of chirality. But some scientists are looking for fundamental reasons for the choice of chirality on Earth, such as the weak interaction, which might violate the symmetry of matter on a very fundamental level. If so, then the weak interaction is directly related to the existence of Higgs boson, which itself has never been observed but still is a major goal of the Large Hadron Collider at CERN. In this case, the emergence of life from the observed predominant chirality is based on the most fundamental properties of our universe.

The viscoelastic approach helps eliminate contradiction between what is highly discussed among scientists, the deterministic scenario of an initial pre-cell evolution, and the contingency theory. The emergence of viscoelasticity is deterministic, but the emergence of something like RNA is still a probabilistic event with highly enhanced probability in viscoelastic media.

Thus, the proposed viscoelastic hypothesis of the initial phase of life, though not absolutely complete and needs a great number of refinements, offers the potential to explain some of the effects associated with the transition of nonliving chemicals to the pre-cell state of living matter. The same probably relates to chirality as well.

Change in sets of initial conditions on Earth at the time of prebiotic state theoretically may lead to a different form of life. It means that some spectrum of trajectories of many different life forms may exist. Theoretically, the approach suggested in this book may help to specify a range of parameters in initial conditions. Physical-chemical interactions may limit environments that are suitable for life and the possible range of their variation is restricted by only the condition: the possibility of the viscoelastic state of matter within this range. The emergence of viscoelasticity inevitably will lead to life formation. All main events in life formation like self-organization and others could happen, be sustained, and evolve in the presence of viscoelastic media.

The essence of the viscoelastic model is that a prebiotic molecular evolution from the simplest molecules, which were widely available on the early Earth, reacting into more complex chemical structures of oligomers, eventually led the system to the state of viscoelastic gel that helped to evolve the first RNA-like molecules. Viscoelastic feedback loops in such prebiotic systems with bifurcations hysteresis are the major selective evolutionary factor. During the process, viscoelastic phase separation in the somewhat ordered structures of the gel lattices could have an effect on selection of at least some of the complementary molecular strands coming in from the reaction zone pools of relatively random sequences. Viscoelastic phase separation could have been a first mechanism leading to macromolecular synthesis and/or selection.

Nevertheless, researchers at the University of Gothenburg have now come up with a new method to control the chirality in the process of chemical synthesis. In biomolecules like DNA, sugars, lipids, and proteins exist in only one form, either left-handed or right-handed only. But the same does not apply when chiral compounds are created synthetically in the lab. Generally, an equal amount of both enantiomers is produced. Theonitsa Kokoli of the University of Gothenburg's Department of Chemistry has been working with absolute asymmetric synthesis instead, where optical activity is created [42]. This was considered impossible by many organic chemists before. He has used reactions, where the two forms have crystallized as separate type crystals, which in itself is fairly unusual. The product that was formed after the reactions comprised just one enantiomer. If such a thing could happen in the crystallization of supramolecular structures in a viscoelastic medium, the chirality problem can be solved successfully.

Another approach to solving the problem of chirality was made by French teams of scientists led by Louis dHendecourt, CNRS senior researcher at the Institute d'astrophysique spatiale (Université Paris-Sud 11 / CNRS) [40]. Researchers for the first time obtained an excess of left-handed molecules (and then an excess of right-handed ones) under conditions that reproduce those found in interstellar space. This result, therefore, supports the hypothesis that the asymmetry of biological molecules on Earth has a cosmic origin.

However, I no longer want to go into the issue of chirality due to the existence of many fundamental difficulties. This central problem of biology requires for its solution the involvement of the efforts of various branches of science: mathematical and statistical physics, the theory of nonlinear processes, kinetic physical chemistry, etc.

16
Four-letter code

It is also important to note that the above arguments and approaches do not guarantee the evolution of the primordial soup into RNA- or DNA-coded life. The approach that is described here only explains that the initial production of relatively long molecules that form gel matrices inevitably leads to evolution of structures based on them, provided there is enough energy and matter for chemical reactions. It seems the life that we see based on the four-letter code is in fact the result of such evolution. This approach explains that in order to form the first nucleotide, there is no need to wait for an accidental combination from all the forming monomers. In addition, the creation of an artificial biopolymer matrix of DNA molecules has led to interesting results, which I think to some extent confirm the participation of viscoelasticity in the cybernetics of life. In a study [43], the researcher showed that DNA-based neural networks—a viscoelastic soup of interacting molecules of biopolymers—demonstrate the ability to take an incomplete pattern and figure out what it might represent; this is one of the brain's unique features. Such biopolymer systems with basic, decision-making capabilities could have powerful applications in medicine, chemistry, and biological research. Theoretically, in the future, they could operate within cells, helping to answer fundamental biological questions, or diagnose a disease. Biochemical processes that can intelligently respond to the presence of other molecules could allow engineers to develop increasingly complex chemicals or build new kinds of structures, molecule by molecule. The intelligence of such systems in my opinion is determined, at least partially, by their viscoelastic properties.

From what we have discussed, we can conclude that the resulting life form may not be the only possibility. Of course, if we consider that our cybernetic system of mutating molecules is aiming toward the "closest" stationary state possible at each iteration, then, theoretically, other life forms can be created. That's why the resulting final stationary state determined by the four-letter genetic code is probably also affected by certain initial physical conditions: the chemical composition of the proto-solar cloud out of which the solar system was born; the local composition of the proto-planetary disc out of which Earth was born; the level of solar radiation or radiation from the Milky Way in this region of the solar system; atmospheric pressure; mineral content and viscosity of water; and so on. If the external conditions were different, then this possibly could have led to a different final stationary state with some sort of N-letter genetic code with an N value other than four. It may be that this alternative possibility—the fact that mutations and evolution are possible not only in RNA or DNA systems—can be illustrated by one study [44]. This study showed that prions (large protein molecules) can have many mutations and that they can bring about evolutionary adaptations such as drug resistance, which was previously known to only occur in bacteria and viruses. The prions are able to adapt and survive in a new host environment, producing distinct self-perpetuating structural mutations that provide a clear evolutionary advantage. Because in the stated process one of the main roles is played by viscoelasticity, then, theoretically, it is possible to assume the existence of different kinds of life based on different chemical elements. The only necessary condition is that the emergence of viscoelastic matrices must be possible.

Nevertheless, it is also important to note that according to the data from one study [45], the modern-type system for the double-stranded DNA replication evolved independently twice: in the bacterial and archaea/eukaryotic lineages of cells. In this case we must admit that the convergence of evolution to the four-letter coding system is very strong. However, it is quite possible that during evolution of any kind, a certain first stationary state is reached that allows for survival of the species, but not with the lowest possible level of entropy. This is probably why our life expectancy does not greatly exceed the age of sexual maturity. Here the evolutionary mechanism is not remotely interested in achieving the next, hypothetically possible stationary states, in which we could live much longer. Contrary to popular belief, maybe evolution does not select the fittest from the available, but stops its selection at the fittest from the accidental. Here the term "accidental" implies the stationary states that have not only been realized due to the inherent properties of live systems, but also due to chance factors of the external environment. That is, as soon as a new species appears, it takes up all or most of the space and does not allow similar species to emerge. This assumption may not be that absurd. If a mutation has assumed its rightful position as a life form, then the rules of the game change for every consequent mutation that could potentially be even better. And the "consequent" mutation is in a less advantageous position by definition, as the "previous" mutation has had free space to allow it to spread and develop. Here the advantages of the new mutation may not be sufficient to outcompete the previous species and survive.

So in this hypothetical situation, evolution will resume when the external conditions change; however, it will move in a slightly different direction. We can say that the evolutionary trajectory has changed. The same can be said in the case of some species diverging. If one branch fortuitously gained an advantage and colonized the habitat, then, potentially, the more advanced branch may not survive. In such a scenario, continued evolution may not translate into ever-increasing fitness. The evolutionary variety is very well represented by the unique properties of animals and plants that evolved in isolation—for example, in Australia or the Galapagos Islands. Also, this idea is supported by a recent study [46] and demonstrates that the struggle for survival is stronger between more closely related species than those distantly related. The authors found that species extinction is happening more frequently and more rapidly between species of microorganisms that were more closely related. While evolutionists generally accept the premise, the study contains the most direct experimental evidence yet to validate it. Darwin's idea of "survival of the fittest" has been called into question as well in an article published not long ago [47]. Researchers challenged this old paradigm by showing that biodiversity may evolve where previously thought impossible. Authors have been watching hundreds of generations of bacterial evolution, about three thousand years in human terms. It had been believed that the genome of only the fittest bacteria would be left, but that wasn't their finding. The experiment generated unexpected genetic diversity in which both the fit and the unfit bacterial cultures coexist indefinitely. Also, in an another slightly older study[468] published in the Royal Society journal *Biology Letters*, researchers provide further evidence that random genetic mutations over millions of years may also play a powerful role. The authors conclude that evolution should not necessarily be called "survival of the fittest" but rather "survival of the luckiest," which is closer to the aforementioned principle of "survival of the fittest from the accidentals."

Researchers at the University of Texas at Austin, led by Matthew Cowperthwaite, show in their computer models that life may not always be optimal and natural selection may not produce the best organisms [48].The team developed numerical models of RNA molecules evolving by mutation and natural selection. Their computer models show that the evolution of optimal organisms often requires more than one mutation, in fact a long chain of interacting mutations, each arising by chance and surviving natural selection. As the authors explain, "some traits are easy to evolve—formed by many different combinations of mutations. Others are hard to evolve—made from an unlikely genetic recipe. Evolution gives us the easy ones, even when they are not the best" [49]. Indeed, what the authors of this study call "easy ones" I call "the fittest from the accidentals."

However, perhaps nature has found a method to circumvent the above problems of evolution through the mechanisms of simultaneous (or cooperative) multiple mutations [50]. The study already demonstrated such a possibility that if mutations can happen cooperatively in twos, threes, or even more, then cells could make large evolutionary leaps through "fitness valley" and reach a different "fitness state" by acquiring multiple mutations simultaneously. It is most likely that the evolutionary mechanism is "interested" in the maximal speed of evolution.

Returning to the nonlinear dynamics of our gel system, it should be noted that a key factor in the evolution of our primitive gel is a high frequency of mutations. Probably similar to what is described above, high frequencies or simultaneous multiple mutations mechanism in the primordial gel have led to its fast evolution. The final result of the cybernetic evolution of the live gel was possibly the creation of first primitive RNA molecules, which relatively quickly led to the emergence of the simplest DNA in primitive cells. So the "live ocean" in Stanislaw Lem's "Solaris" is a theoretically possible scenario for the evolution of a pre-live gel in the ocean of a planet with suitable environmental conditions. It is important to note that it is most likely that the primordial gel will be localized on the bottom of the ocean or coastal-area ponds, in both cases filling up cracks and porous rocks of volcanic origin, for example, for the best protection from external influences and the abundance of minerals and chemicals in young volcanic rocks.

The formation of this primitive gel is a fact of Gaia emergence, metabolizing by mutations of a partly viscoelastic substance, involving in its own metabolism the planetary environment. While of course it is not possible to deny the first cells could have arisen very early in the history of Gaia. In this case, the whole process of viscoelastic evolution held inside the cells. But this does not affect the fundamental feasibility of the proposed concept in this book.

17
Belousov-Zhabotinsky gels

It seems once again necessary to emphasize the fundamentality of the viscoelasticity of biological objects, from molecules to supramolecular structures. Currently, the viscoelasticity of the macromolecules in biology is generally treated as a trivial consequence of the presence of very long molecules. In this book the role of viscoelasticity is treated as a fundamental, necessary, and sufficient property of biopolymers for the process of evolution and maintenance of the observed state of living matter. In the hypothetical situation, if macromolecules would not have the viscoelasticity, it seems that the living state of matter at least in the form of Gaia could not arise and exist. And this is in spite of all kinds of patterns, the researchers found in the analysis of Turing or other kinds of patterns in the diffusion-reaction mixture.

These seemingly speculative ideas about early evolution have, in my opinion, a heuristic value for the understanding of living systems. It is possible that the "explosive" cybernetic evolution during the formation of the living gel laid a variety of molecular mechanisms that provide the great variety of life that we have on Earth. I think it is an appropriate "big bang" analogy at the beginning of time. Physicists also quite reasonably suggest that the properties of particles, fields, and perhaps the laws of interaction that occurred at the time of the formation of the universe determined its modern structure.

Besides the ideas outlined above, the origin of life does not require any exobiological causes, such as panspermia, or the idea of God. Moreover, this approach, theoretically, allows the possibility of the "other" type of evolution under different external conditions or on the basis of some other chemical elements, leading to different metabolic pathways and life forms. And you must also specify that the proposed approach opens up the possibility of a life, so to speak, per se (lat.), due to the existing physical-chemical diversity of the universe, its laws, and the nonlinear mechanical properties of matter.

Thus, the viscoelastic hysteresis of bifurcations apparently played a key role as the primary mechanism of delayed feedback (relaxation time), which creates homeostasis protozoa and gave it an observable steady state. Dynamic equilibrium in the homeostasis of the body, to a large extent, was determined by the presence of viscoelastic properties in all structures of any organism, starting from macromolecules and organelles, the cytoplasm and cell membrane, and ending with elasticity of organs, tissues, bones, and a non-Newtonian rheology of blood and lymph. In short, all that is in the body.

In linear viscous media reversible self-sustainable reactions have a self-oscillating nature, such as the classical Belousov-Zhabotinsky (BZ) reaction with a periodic change of color in a homogeneous solution. In viscoelastic polymer media, if we imagine that the BZ reaction depends on the viscoelastic state of macromolecules, the situation is slightly different. Such a self-oscillating reaction will have the character of propagating waves at a certain speed. The propagation velocity would be determined by the characteristic relaxation time of the solution.

If in such a *in effigo* (in mind) BZ experiment we would have a mixture of reactive polymer solutions with different relaxation times, the waves would have a reaction fractal structure: on the borders of waves (in the areas of color change of the solution), there would be areas of shorter waves, corresponding to the smaller relaxation times in the spectrum of relaxation times of the mixture. I think that such a situation is realized in all the systems of the body, where in vivo there are so many molecules of polymers in an even greater variety of conditions and network interactions with a virtually endless range of relaxation times.

In the simplest case the BZ reaction is an elementary nonlinear (the model for living systems) self-sustained process characterized by a periodic change in the color of the solution. But in living cells and mediums involved millions of proteins with different lengths and shapes of macromolecules and therefore set the relaxation times in such system, from a practical point of view, is endless. That is, at the same time in a real cell there is a huge set of fractal wave reactions determined by the feedback loop function with so-called 'fading memory' (i.e., the influence of distant in a time events is less than recent). In view of the boundary conditions and the fact that the cell stably exists in time, it can be assumed that in the space of its metabolism, the superposition of standing waves of reactions occurs that is observed in vivo as the heterogeneity of the cell structure.

Most people look at a living cell as something with a more or less permanent structure. But they are badly mistaken. In the cell, there is nothing permanent. All parts of the cell are the supramolecular structures formed from polymer or lipid molecules that are unstable and are constantly in the process of association-dissociation. The prevalence of the association process is facilitated by the existence of relaxation time of the viscoelastic phase in the process of dissociation of the supramolecular structures. That is, if the processes of molecular association in supramolecular structures from the cytosol, nucleosol, in membranes, etc., occur relatively quickly, the dissociation of structures usually takes more time. The reason for this hysteretic behavior is the difficulty in the movement of polymer molecules or their end caps in supramolecular ordered structures, similar to liquid crystals, due to the interaction of neighboring molecules in the viscoelastic phase separation state. You can also consider the structure of the cell as a superposition of many oscillatory chemical reactions, such as the BZ reactions in a viscoelastic medium with an exceptionally wide range of relaxation times. Each relaxation time (or, to be more precise, narrow parts of their spectrum) is responsible for all cellular structures and determines the fractal dimensional state of the cells in the phase space.

Such supramolecular structures are generally very small—much smaller than any part of a functioning cell—organelles. The transitory process of the micro-phase of supramolecular structures to bigger structures can take place if the local concentration of supramolecular structures exceeds a certain limit necessary for phase transition. Concentration growth in this locality is a narrowing of the ensembles of the probabilities of the states of particles due to the high concentration of supramolecular organized micro-phase. If transitions from small supramolecular structures to organelles like supramolecular macro-phase and back have hysteresis character, then the micro-phase system may jump to the new macro-phase state.

In fairness it should be noted that in some polymers, for kinetic reasons, the association of supramolecular structures may take a longer time than their dissociation. Thermal hysteresis and the stability of agarose-water gelling systems were studied in [53] at different microscopic levels. Researchers found the two distinct orders (molecular and supramolecular), i.e., branched double helices of polysaccharide chains and bundles of associated double helices. The two types of order appear to be lost through two distinguishable transitions. The supramolecular order disappears gradually over a temperature span of about 50 degrees when the temperature of the gel is increased. Over the same temperature span, it appears reversibly restorable without hysteresis, although with long kinetics. The molecular order instead appears and disappears through much sharper, largely hysteretic transitions with much quicker kinetics. Mimicking of thermal hysteresis in supramolecular ordering appears to be only a consequence of the fact that this ordering requires the existence of molecularly ordered "blocks" to occur. The author's interest in the systems of this type lies in their being recognized as model systems of a variety of biologically organized structures.

A good example of autowave processes is self-oscillating polymer gels, which are materials that continuously change back and forth between different states—such as color or size—without provocation from external stimuli. As we can see from experimental [54] and computational modeling [55] gradients in cross-link density in a polymer matrix—a system very similar to developing polymer/oligomer matrix with spatial distribution of viscoelasticity—lead to the bending and self-propelled motion of active gels. Oscillating polymer gels undergoing the Belousov-Zhabotinsky (BZ) [54] reaction provide an ideal model for probing the interaction between chemical energy and mechanical action. The BZ reaction in the gel is converting chemical oscillation to mechanical oscillation of the polymer matrix. You can watch it in a video [56]; these self-sustained pulsations are caused by the BZ chemical reaction's interaction with the polymer matrix. Without any outside influence, oscillation from this chemical reaction can develop within the material or if the gel size is small enough, cause the entire gel itself to pulsate mechanically for several hours. And vice versa, in such systems mechanical stresses in a matrix could be converted to chemical changes in mechanotransduction processes.

Similar results were obtained in recent research [57, 58] where the author introduced an autonomous gel as an actuating device transmitting chemical energy to mechanic motion. The polymer gels prepared had cyclic chemical reaction networks. It was shown that with a cyclic reaction, the polymer gels generate periodical motion. The periodic motion of the gel is produced by the chemical energy of the oscillatory BZ reaction. The author had succeeded in making the synthetic polymer gel move autonomously like a living organism and even in conveying the object by utilizing the peristaltic motion of the polyacrylamide gel. Although the gel is completely composed of synthetic polymer, it shows autonomous motion as if it were alive. These results again emphasize that the basis for the functioning of living organisms is the viscoelastic properties themselves.

But gel oscillations play a very important role in the real cells. So, for example, they are known to be associated with successful progress of an embryo development. Professor Magdalena Zernicka-Goetz of the University of Cambridge led a team of researchers to look for ways to assess fertilized embryos more effectively, allowing fewer embryos to be implanted. In her experiments on animals, she found that when a sperm entered an egg, the egg's jellylike innards would start to pulsate soon afterward [59]. The oscillations seen by researchers in the egg's cytoplasm—the jellylike liquid inside the cell—are caused by the influx of calcium ions after an egg is fertilized [60].

This temporal control in just fertilized embryos is executed by modulating the force balance between two states of cell cytoplasm, biopolymer matrix strength and through feedback loops based on signaling cascades of cytosolic calcium concentration. The cytosolic calcium concentration inside the cell is increased initially by probably an influx of calcium ions with or in presence of sperm and decreased by the binding of calcium ions with the biopolymer matrix of the cytoplasm of fertilized egg.

A feedback loop emerges in a signaling network that connects to itself, forming a circuit or loop. An initial fluctuation of one component will propagate through the loop until it feeds back unto itself and amplifies or reduces this initial fluctuation. When the intensities of positive feedback (amplification) and negative feedback (suppression) are balanced, a temporal state such as regular oscillations of the parameters of the system can emerge.

In the embryo case observed in [60] the binding of calcium ions with the biopolymer molecules of the matrix may some way change their viscoelastic properties and consequently the chemo-mechanical state of cytoplasm gel—mesh size. This leads to changing volume taken by the gel. But when some polymer molecules or their parts elongate, bonds of calcium ions with polymers break down and the gel returns to a previous state, but the concentration of ions in cytosol in a mesh space increases and the process starts again. Such type of feedback loop, I believe, is involved in the emergence of embryo pulsations as presented in research [60]. Obviously, pulsations in human embryos are similar to those seen in animal eggs, because of similarities in the embryos' biochemical properties and size.

As well, a team led by the Princeton professor of molecular biology Ned Wingreen reported recently in the journal *PLoS Computational Biology* [61] interesting results that, contrary to the observed fact that embryonic cells develop in synchrony, they are, in computational models, prone to descend into chaos. From the beginning, embryonic cell cycles are initiated by a wave of calcium ions that emanates from the fertilization site and prompts the embryo's cells to divide and duplicate—to oscillate, in biological terms. Wingreen and his colleagues found, however, that the natural spread of oscillation in computational models is unstable and would result in an erratic patchwork of missed and incomplete cell divisions.

But then the question arises: If *in silico* there's potential for chaos, how does the embryonic system avoid it and get into the synchronized state *in vivo*? To get an observed *in vivo* outcome, oscillations should be arranged throughout the embryo in the stable wave's pattern.

Computational simulation produced the picture that the moving calcium ions wave is known to spark cell changes as a synchronizer that sets embryonic cells into the synchronized development. But in my opinion, in this case as well as others above, the synchronization achieved through the mechanism the calcium ions influx, which are sparking propagating in embryo changes in the state of viscoelastic gels in the cells and throughout this viscoelastic feedback loop is synchronizing embryo. As it should be in typical BZ reaction, the stable waves of calcium emerge with characteristics (amplitude and frequency) throughout the embryo depending solely on the viscoelastic chemo-mechanical properties of the cytoplasmic gels and probably membranes of the cells.

The viscoelastic approach is a good framework to explain the oscillations observed in the life of the egg [61] and nonlife systems with the BZ reaction in an artificial polymer gel in [56]. Of course particular classes of feedback loops could present very distinctive behaviors such as a color changing wave, or the rate of oscillations change in the gel, etc., upon the influencing of an endless number of parameters in real gels. But it is important, as we see from the results [59 - 61], that from the very first seconds of any life, including our own, viscoelasticity plays a crucial role.

A good example of the importance of BZ reactions is cell mitosis. According to the description of mitosis in one study [63], microtubules extend from one of two spindle poles on either side of the cell and attempt to latch onto the duplicated chromosomes. In addition to microtubules from both spindle poles that attach to the chromosomes, astral microtubules that are connected to the cell cortex—a protein layer lining the cell membrane—pull the spindle poles back and forth within the cell until the spindle and chromosomes align down the center axis of the cell. Then the microtubules tear the duplicated chromosomes in half. The process of mitosis is extremely important for the cell. Gaining or losing a chromosome during mitosis may lead to cell disorders and to a whole body of diseases like diabetes, cancer, etc. Authors [63] noticed that when the spindle oscillates toward the cell's center, a partial halo of the protein dynein lines the cell cortex on the side farther away from the spindle. As the spindle swings to the left, dynein appears on the right, but when the spindle swing to the right, dynein vanishes and reappears on the left side. In this process dynein is anchored to the cell cortex by a complex that includes the protein called leucine-glycine-asparagine-enriched (LGN) protein. The stationary dynein acts as a winch to pull on the spindle pole, and the microtubules and chromosomes attached to it, toward the cell cortex.

Researchers found that when a spindle pole comes within close proximity to the cell cortex, a certain signal protein emanates from the spindle pole, knocking dynein off LGN and the cell cortex, stopping the spindle pole's forward motion, and freeing dynein to move through cytosol to the opposite side of the cell. These oscillations continue with decreasing amplitude until the spindle settles along the cell's center axis. Researchers also noticed that a layer of LGN extends all around the cell cortex, except in the areas that are closest to the chromosomes. As the chromosomes swing back and forth, the cortex area cleared of LGN changes in response. Because dynein needs to anchor to LGN, this cleared area ensures that dynein can only attach and pull to the right and left of the aligning chromosomes, rather than from above and below. Thus, we see that the most fundamental process in the life of the cell—the mechanism of the spindle settle in mitosis—is controlled through the mechanical forces originating from the BZ type chemo-mechanical reactions in macromolecular viscoelastic media of binding the dynein on/off LGN and both of them on/off the cell cortex.

The important role of autowave reactions in the processes of cell division was noted in one study [63]. According to this research, initially mitosis cells place their nucleus in the center of the cell, and then bundles of microtubules extend from the nucleus to either end of these cells. In the process of autowave reaction, the microtubules push against the end of the cell, thus pushing the nucleus away from the end of the cell. The likelihood of a microtubule bundle reaching the end of the cell—and thus having the opportunity to push against the end of the cell—drops off as the distance from the nucleus to the end of the cell increases. Thus, on average, microtubules will push more if the nucleus is close to the end of the cell, and the nucleus will be pushed back into the center of the cell.

It is also important to note that the oscillations in the embryo have a unique role in the differentiation of cells—gaining their specialization. No differentiated cells of the embryo as it grows are in different conditions. It is clear that the cells within the embryo and the outside, due to different boundary conditions, have different frequency pulsations. I even do not care and do not know what these boundary conditions are on the embryo at this time. They can be quite rigid boundaries, such as, for example, the uterine wall, or, on the contrary, they can be free boundary conditions, such as when the embryo floats in a fetal fluid. The main thing that defines a border is a different mechanical state of the cells in the outer part of the embryo and in the interior.

Thanks to the autowave mechanochemical process, ripple matrix membranes, cytoplasm, and the skeleton of cells and nuclei (usually called mechanochemical transduction), the wave of redox potential reaches the cell's DNA and causes resonant mechanical pulsations in some sites of it.. This in turn provokes the loosening of chromatin at these sites and the expression of the corresponding genes. As the frequency and amplitude of the fluctuations on the outer edge of the embryo differs from that of the inside, the resonance conditions and gene expression in cells at the border and within the embryo will be different. So I see the mechanism of the initial differentiation of cells in the embryo.

18
Universal chaos, fading memory, and role of viscoelasticity in body functioning

Many biological processes, such as embryonic and organ development or tumor growth, involve the remodeling of tissues by cell division and cell death or apoptosis. For many years, emphasis has been put on the regulation of growth by signaling pathways such as growth factors and their genetic control [64]. Recently, however, the importance of the mechanical viscoelastic properties of tissues has been realized. It is already well known that during the development of embryos the expression of the genes can be strongly modified by the application of external forces that change the local mechanical stresses acting on the cells in the growing organism. At embryonic and other stages of development the spatial distribution of mechanical stresses also seems to play a key role in controlling the pattern of gene expression. As in tumor progression, the gene expression pattern is related to the stress distribution and the cell position in the tumor.

Earlier I noted that the process of renormalization of chaos to a new state in a viscoelastic medium is responsible for the molecular evolution and origin of life. The same process is responsible for the evolution of supramolecular structures and for the evo-devo (evolutionary development) work of the genome of modern organisms. The genome of any multicellular organism is a heterogeneous 3-D matrix of chromatin in the cell nucleus. Transition states that occur in the matrix happen through chaotization of its structure and restructuring the matrix to a new state. These affect the genome through the genes' expression status, the rate of replication of various RNA, etc. In addition, be aware that the whole cell activity is the work of auto-oscillating matrices. All transitions between their states are of a hysteresis nature and are driven by renormalization. This, in other words, means that the viscoelastic gel matrix in living systems is always "on the edge of collapse." Any alteration of the matrix is accompanied by chaos, i.e., complete or partial destruction of some matrix area and its transition to a new state. And with it is the entire matrix.

Thus, the existence of living matter from the first moment to any point in the present day for the last four billion years is determined by the same universal process of transition from chaos to order of the new structure. Within each cell of the body this happens millions or billions of times per second. This process is the basis of life. It explains why life is not an exceptional fluctuation, but a natural physical process that is subject to the laws of thermodynamics. I especially emphasize here the thermodynamic nature of the life process, because a single fluctuation is not a thermodynamic process. By definition, only one all-powerful God does not obey the laws of thermodynamics. But we are not looking to that hypothesis in this book.

However, the versatility of the approach based on the viscoelastic renormalization is also confirmed by the analysis of the processes of development of tissues and organs. During the formation of tissues, cells organize collectively by cell division and apoptosis. According to one study [64], the multicellular dynamics of such systems is influenced by mechanical conditions and the process of chaotization (fluidization), which both give rise to tissue rearrangements to new order. The process of dynamic reorganization of elastic tissues leads to chaotization, which manifests itself as liquid-like viscoelastic behavior with well-defined shear and bulk viscosities on longtime scales and in the vicinity of the homeostatic state. Chaotic noise in the tissue appears due to chaotic cells division, cell death, and cell shape fluctuations. A unique consequence of these cellular reorganizations in the vicinity of the homeostatic state is the absence of a compression modulus. Renormalization of the chaotic noise in the system leads to a new organization of tissue through the coordinated movement of cells observed experimentally in the development of embryos and tumors. This process is continuous in developing tissues upon reaching certain limits of elastic deformation in the rubbery tissue.

In evo-devo processes, the main source of chaos and convergence to a new state are processes in the DNA of cells, tissues, and embryos. Chaotization of the genome happens under external mechanochemical signals from the surrounding tissue. Further convergence of the genome to a new epigenetic state determines the work and mobility, and the new configuration of the tissues after reformatting.

We should also mention that there is the generation of negative entropy in such systems. In reviewing the processes with the fading memory (non-Markovian processes) characteristic for viscoelastic media, accounts for the prehistoric naturally introduced an additional characteristic—negative entropy as a measure of the emergence of order and complexity of the structure of processes in chemical systems.

The Markov process is a random process for which there is a known state of the system at any time. Its further evolution is not dependent on the state of the system in the past related to this point. Obviously, the Markov processes are not suitable for describing polymer pre-biologic systems. The non-Markovian nature of things is present in such phenomena as the dependence of stresses in viscoelastic liquids, not only on the current value of the rate of deformation but also on the characteristics of its history—the spectrum of relaxation times of the system. The non-Markovian approach proved extremely fruitful for the polymeric non-Newtonian fluids, in particular, and to describe processes with hysteresis. When oscillatory chemical reactions happen, such as BZ reaction in the polymer gel as the temporal and spatial variation of viscoelastic properties of the gel emerge, there is a huge variety of increasingly complicated structures appearing. The characteristic features of non-Markovian processes is the existence of certain stable spatial and temporal cycles that manifest themselves in BZ reaction in the gel, as sustained mechanical oscillation.

That the processes of mechanical pulsations initiated by temporal/spatial changes in the viscoelastic matrix take one of the dominant places in the mechanism of the cell functioning also follows from the results of some experimental studies. For example, the process of reconstitution of actin matrix that is actively set under stress by molecular motors (molecules that transduce chemical energy into a directed force) has been studied in research [65]. The results of the study show that in an active and cross-linked matrix pulsatile collective modes develop, which depend critically on the interaction strength of the molecular motor and filaments. This type of interaction in some sense is very similar to BZ reaction when the dynamic process of movement of molecular motors through the gel is changing the viscoelasticity of the gel locally. In turn, the feedback mechanism leads to changes in the molecular motors' dynamic. Indeed the force exertion inside the dynamically cross-linked actin matrix of cells is expected to induce a force rate dependent on rupturing of the cross-linking points and thus the locally induced, through the dynamic matrix rearrangements, the dynamic changes of the viscoelastic environment, will also affect and modulate the molecular motors' dynamic. Thereby, it would be conceivable that the dynamic viscoelasticity of the matrix enables the appearance of collective pulsatile modes. These pulsations can, theoretically, do both to slow passage of a molecular motor with the cargo and to accelerate it sometimes, due to the effect of viscoelastic peristaltic pumping. Especially, this effect may be noticeable for long clusters such as the motor-signal protein.

One of very important process of pulsations is the work of our heart. Abnormal heart rhythms—arrhythmias—are killers. They strike without warning, causing sudden cardiac death, which accounts for about 10 percent of all deaths in the United States. This is more than 250,000 people. "The current antiarrhythmic drugs do not prolong life," said Björn Knollmann, associate professor of medicine and pharmacology and the senior author of the current report. "There's a large need for new approaches to antiarrhythmic therapy" [66]. In their quest to understand how irregular heart rhythms arise—as a way to find new molecular targets for treatment—Knollmann and his colleagues have focused on the role of calcium inside heart muscle cells. Calcium ions are central to the contractile cycle not only in embryos, as we saw above, but for heart tissue as well. Electrical impulses based in calcium ions regularly circulate through cardiac tissue and cause the heart's muscle fibers to contract. The calcium ions interact with proteins called troponins, which are part of the heart tissue contractile filament. The interaction of ions with troponins regulates filament contraction. In a healthy heart, these electrical impulses provide smooth travel for contraction-relaxation waves through tissue. But mutated troponin had been linked to inherited forms of hypertrophic cardiomyopathy (HCM), which carries a high risk of sudden cardiac death. HCM is perhaps most famous as a cause of sudden cardiac death in young athletes, but it can affect individuals of any age [66].

The researchers examined the heart rhythms of mice expressing various troponin mutants that cause HCM and showed that the mice develop ventricular tachycardia (a particular arrhythmia). The risk for this arrhythmia was directly related to the degree of calcium sensitization caused by the troponin mutation: the higher the calcium sensitivity, the greater the arrhythmia risk. Troponin mutations associated with HCM increase the sensitivity of the troponins to calcium—they bind calcium ions more readily, which activates the filaments more easily and results in the increase in mechano-responsiveness and stronger contractions. Increased calcium ion sensitivity has also been found in acquired heart diseases, such as heart failure, that have a high incidence of sudden cardiac death. Authors [66] proposed that increased filament calcium sensitivity contributes to arrhythmia susceptibility.

In another study [68] on the factors that lead to fatal cardiac rhythms, a team of Canadian researchers has shown that the importance of communication applies to the heart cells. But what does cell communication mean here? It is based on calcium ion signals getting from one cell to another under electrical forces through the cell and intercellular space. The heart cells are electrically excitable cells, maintaining voltage gradients across their membranes by means of metabolically driven ion pumps, which combine with ion channels embedded in the membrane to generate intracellular-versus-extracellular concentration differences of calcium ions.

In the mechano-sensitive ion channels, signals can be produced when forces acting within the viscoelastic lipid bilayer [67] rise to a level sufficient to produce conformational changes in the channels forming proteins and thereby alter their conductivity. In a young person these adaptive properties of membrane ion channels are big enough to pump through the cells surface a large number of ions. But with age and mainly with long periods of incorrect disanthropic diets and lifestyles, due to epigenetical changes of proteins and lipid matrices, the forming channels are losing their flexibility. And when there is a sudden increase of heart activity, ion channels cannot provide enough ions to the cell's communication pathways.

Modulated by the viscoelastic properties of ion channels, electrochemical signals travel from the cell membrane and activate pathways for electrical impulses carried by ions to other cells through intercellular space. The viscoelastic property of heart cells is very important for the adaptation of cells' ability to transfer electrochemical signals in a wide range of intensity of mass transfer.

Equally important is how the ions travel and bind in intercellular space. In their work [68] when the researchers studied chick-embryo cardiac cells grown as a sheet of tissue, in the first two days after this arrangement of cells is created, spiral waves often form in the tissue. When the researchers sprinkled the sheet of cardiac tissue with a drug that impairs communication between the cells—partially interrupt electric signal transfer by calcium ions—they observed that the rotating spiral waves broke up into multiple rotating spirals. This breakup of spiral waves in the two-dimensional sheet can sometimes become troublesome and is believed to be similar to the 3-D electrical patterns that cause human hearts to undergo ventricular fibrillation, a potentially fatal cardiac rhythm that often occurs when the ion communication channel between cells is impaired. Impairment of signaling between cells in this model experiment means that the drug changes the viscoelastic property of the intercellular matrix and the conditions of ion transfer, of binding them and contraction of the matrix. Instead of homogeneous contraction impulses cardiac shit, we have heterogeneous shit, where cycles of contractions have, instead of one, two mechanical relaxation times.

As we see from research cited above, we have three reasons for failure of the heart waves: loss of adaptation of membrane ion channels, troponin mutations changing in binding processes of calcium, and changes in intracellular matrix. All three reasons are emerging by changing solely the viscoelastic properties of cells, intercellular matrix, and filaments. Loss of elasticity of channels in the heart cells and of the intercellular matrix creating new heterogeneous (for electrical impulses) tissue. Coexisting two type of tissue with two different relaxation times of electromechanical impulses leading to the emergence of two propagating front waves in the heart tissue and leads to observed disorder in the heart pulsations. Additional heterogeneity in the tissue—appearance fragmented areas of diseased tissue giving as well new additional waves propagating quite independently and destroying normal coherent pulsations. When neighboring cells have strong signaling and mechanical interaction with one another, electrical waves quickly pass through the tissue unobstructed; when interactions are weak, wave propagation is completely blocked.

At intermediate levels of interaction, electrical waves break up into multiple spiral waves.

Troponin sensitivity to calcium as well changes the relaxation time of the contraction-relaxation processes of the lattice. That is, it leads to stronger heart beats that finely lead to faster destruction of heart tissue. If some part of the cells in the tissue has different troponins, in this case two and more waves in a heart might emerge. How are we to avoid both of these nasty outcomes? Medical researchers are trying to find drugs with calcium desensitizing activity. But inevitably that drug will have huge side effects as with almost all drugs, because there may be situations when some part of tissue needs desensitization for calcium, but another not. It is even more difficult to manage ion channels functioning. Only understanding the effects of calcium ions communication between cells and the influence of viscoelastic properties of matrices provides insights into the electrical malfunctions that are suspected to lead to heart disorders, and may ultimately suggest strategies for avoiding them. I think, personally, that it is almost impossible to find drugs or combinations of them, if two or all three reasons for arrhythmia are acting together. The possibility to get a long-lasting effect is to change your diet [1] and to try to restore epigenetic changes if they have not gone so far. If we can restore the ability of neighboring cells to strongly interact with one another and prevent epimutations of troponin, electrical waves will pass through the tissue normally almost at any age.

Also, it is interesting to note that viscoelastic properties apparently play important roles in the work of our brains. In the work [69] focusing on identifying the mechanisms by which cells sense mechanical stress and transduce it into a biochemical signal, a three-dimensional viscoelastic model of mechanotransduction is developed. Cells are exquisitely sensitive to mechanotransduction, and actively respond through a variety of biological functions, including migration, morphological changes, and alterations in gene expression and protein synthesis. The study found that due to the viscoelasticity of a cell matrix, the time-dependence of the sinusoidal force manifests itself in the stress concentration within the cell. The faster loading rates lead to higher stress concentrations in the cell. Stresses transmitted via cell surface receptors and the intracellular proteins that connect them to the intracellular matrices can also induce conformational changes in them and, as a result, potentially alter matrices binding affinity to signaling and other molecules or their diffusion rates.'

Neuroscientists have long believed that vision is processed in the brain along circuits made up of neurons, similar to the way telephone signals are transferred through separate wires from one station to another. But scientists at Georgetown University Medical Center [70] discovered that visual information is also processed in a different way, like propagating wave packets. They visualized wavelike patterns in the neuron network of the brain cortex using a new method called voltage sensitive dye imaging. A collective pattern emerges from the activities of billions of neurons in the visual areas. The wave packet patterns obviously play an important role in initiating and organizing brain activity not only in visionary processing but in the memory functioning in the brain as well. It means that the brain cortex is processing information by the wave packets spreading in the entire network of neurons in some areas of brain—some sort of parallel computing. It looks like the brain processes all sensory information this way.

The same team uncovered spiraling wave propagation, resembling spiraling patterns in the heart tissue pulsation [66], in animal epilepsy models. Authors think that this multiple spiral pattern in an area of damaged neural tissue can generate disorderly waves that invade healthy brain areas and start a seizure attack. This does mean that disorders such as epilepsy could be viewed not just as mis-wiring in the brain, but as an abnormal electrochemical wave pattern that invades the entire brain network.

The neurons are electrically excitable cells, maintaining voltage gradients across their membranes by means of metabolically driven ion pumps, which combine with ion channels embedded in the membrane to generate intracellular-versus-extracellular concentration differences of ions such as calcium, etc. In the mechano-sensitive ion channels, signals can be produced when forces acting within the viscoelastic lipid bilayer [67] rise to a level sufficient to produce conformational changes in the channels forming proteins and thereby alter their conductivity. Modulated by the viscoelastic properties of ion channels, electrochemical signals travel along the cell axon and activate synaptic connection with other cells when they arrive. I am sure that viscoelastic properties of brain tissue are playing most fundamental role in its work.

19
Fractality of metabolism

In nonlinear systems (strange attractors) elasticity of the media may deprive small (infinitesimal) perturbations from independency, so that the functions that describe their behavior depend on the functions describing the behavior of finite (large-scale) perturbations. That is, in other words, the behavior of the system at the lowest level of the hierarchy depends on the properties of the system at the next higher level of the hierarchy. For example, in the diffusion and reaction of proteins in cells (pores), protein matrix cells are dependent on the properties of the matrix. In such a matrix it is natural that the viscoelastic properties have a significant influence on the state of the phase space of the strange attractor of metabolism.

The virtually infinite (due to the large number of reactions) fractal dimension of the strange attractor metabolism of animals and plants allows its breakup into multiple hierarchically connected strange attractors of different levels of metabolism. For example, the hetero-phase structure of the cell can be divided into many separate phase attractors. Also, in organelles and around them exists their own protein matrices that regulate the speed of transport of signaling molecules.

If we consider the matrix as a relatively low-dimensional strange attractor, and the matrix liquid with signaling molecules as a stochastic set in multidimensional phase space, then their consolidation, because in reality they are consolidated by a cell, is a multidimensional strange attractor.

The viscoelastic mechanism of operation of the strange attractor of metabolism demands, along with dissipation, recognition of elastic properties of tissues and organisms, which determine its adaptation abilities as a viscoelastic deflection point of attraction in the phase space.

I think that the manifestation of viscoelastic properties is one of the most fundamental differences between living and nonliving matter. Indeed, all the dissipative structures of inanimate nature only exist under a continuous flow of energy and are stable only for so-called small perturbations. The main feature of living matter is the high resistance to longtime finite-size perturbations of environmental conditions. If organisms have existed only as a classical dissipative structure, changes in the environment would lead to their almost immediate death with a slight delay due to inertial processes. That is because in the absence of a viscoelastic phase, i.e., the polymerization of the molecules and polymer matrices, macro-viscosity of intracellular fluid would be much less than that observed. The characteristic times of not all, but of many biochemical reactions in the body cells are relatively small, and any fluctuations of matter and energy supply to the cell, due to the insignificance of the inertial effects at low viscosity, would lead to the destruction of their structure.

There are cases of direct influence of viscoelasticity of the cells on the function of various organs. For example, one study [80] made the comparison of actin matrix cross-linked with mutant and wild-type proteins in kidney cells. Researchers found that altered viscoelastic properties of cross-linked actin networks contribute to the phenotypic changes in diseased kidneys. It is interesting that the study found that mutations have a drastic effect on the properties of in vitro networks at shear frequencies around 1 Hz, close to the human heart rate. In vivo, the pathological consequences of the mutant cross-linker are most apparent in cells that wrap around the capillaries in kidney, which are subject to substantial shear stresses resulting from dynamic capillary pressure. Intriguingly, a specific evolution of these cells' viscoelasticity is that they have the relaxation time related to the oscillation frequency of the heart. Such fine tuning of the mechanics of evolution of these cells clearly demonstrates the critical importance of the viscoelastic properties of the cells to the functional properties of complex organisms.

It should be noted also that the viscoelasticity of living systems is manifested in many different ways. A good illustration of the ability to influence the elastic properties of macromolecules in the course of the reaction in the cell is the existence of a fractal dualism viscosity of solutions of proteins prone to classify. This dualism or dynamic asymmetry of diffusion is shown, as in the macroscale, a well-known phenomenon of the nonlinear dependence macro-viscosity of polymer solutions of the shear rate, and at the nanoscale, lower (by five to six orders of magnitude lower) nano-viscosity cytoplasm of the cell compared to macro-viscosity. The nano-viscosity phenomenon was discovered more than half a century ago, in the study of the sedimentation of small particles in ultracentrifuge. Surprisingly, it was found that nano-sized objects move in polymer solutions as if the viscosity of the fluid is tens of thousands times smaller than that recorded by viscometers. In [20, 81] have recently shown that each heterogeneous hydrodynamic media and, in particular, polymer solutions have a characteristic length scale of molecules, in which there is a transition from nano-viscosity to macro-viscosity. This scale length of the molecule is determined by the scale size of the objects of the polymer solution.

So if, for example, the effective size of the coils of macromolecules is around 10 nanometers (nm), then every other particle of the molecules of the polymer, which is bigger than 10 nm, while moving in the solution undergoes macro-viscosity. If it is less, then it undergoes nano-viscosity. The transition from nano-viscosity to macro-viscosity around the characteristic scale is exponential, i.e., very sharp. For example, in [82] it was shown that the anomaly of viscosity (diffusion), i.e., the difference between macro-viscosity (macro-diffusion) and nano-viscosity (nano-diffusion) is significantly reduced if the intracellular environment is under osmotic stress—diluted with water.

It is also known [72] that the diffusion of "small" molecules, such as carbon monoxide in coils of protein macromolecules, occurs at relatively low activation energy. That is, the situation is such that molecules of carbon monoxide very quickly pass through the deformable "matrix" mesh of the coil. The high degree of deformation provides the flexibility of the polymer chains sites (conformational freedom), forming a mesh. But as soon as the size of the "small" molecules increases, the rate of diffusion is reduced by orders of magnitude. In principle, the consideration of such diffusion in the polymer gel matrix to account for the viscoelastic properties of the matrix can introduce the concept analogous to that of the Kuhn segment (relative to the moving parts of polymer) for polymer solutions and melts. Then the number of Kuhn segments between the nodes of the matrix will determine its flexibility, and thus the ability for elastic deformation. This ability for elastic deformation of the matrix cells will in turn determine the character of the diffusion signaling protein molecules and clusters.

The dualism of the viscous properties of polymer solutions, in other words the existence of the dynamic scaling asymmetry of diffusion processes, has a key influence on many processes in organisms, such as the rate of chemical transfer within cells and organelles. In the vicinity of the cell nucleus there is the highly viscous medium containing a high density of macromolecules. Due to the existence nano-viscosity, for small protein molecules the diffusion rate is not very different from their diffusion rate in the pure solvent, in this case water. The high rate of diffusion of relatively small proteins through the matrix of large macromolecules provides a process functioning of the cell. Moreover, the results of [83] demonstrate that the changes in the properties of proteins in the nuclear matrix of the cells induce, through the influence on the properties of the matrix, changes in cell growth and cell division.

The viscoelastic approach allows us to consider visco-elastic matrices of the body as a kind of viscoelastic reservoir of energy, helping to damp the fluctuations in the intensity of metabolism in the characteristic time scale of processes of cells due to the presence of the spectrum of relaxation times of the protein matrix cell. Or, as an equivalent, one can speak about the body as reservoir of negative entropy. Often emphasizing the dominant role of entropy, one can say the elasticity of polymers has an entropic nature, and each polymer molecule is an entropic spring. Viscoelasticity has the most significant influence on the diffusion of proteins close to the characteristic size, which are not enough in the cells and that perform, no doubt, the same critical functions, as well as all the other components of the cell. As for the cell as a whole, and of the behavior of cell clusters, the study [73], based on *in silico* modeling of the viscoelastic properties of the cells, it is shown that the biomechanical viscoelastic properties of cells determine the full range of interaction and communication between cells.

Conclusion

The point of view I present in this book is not from any special views on the processes of the origin of life. How my arguments may seem to differ from what other researchers of this problem are saying is that my approach is strictly based on the physical chemistry and the nonlinear dynamics of the processes. My arguments about the details of the mechanism of early evolution of living matter is not so much based on mathematical analysis, but on some well-known and "obvious" facts observed experimentally and theoretically (mathematically) in the works specified in the bibliography. However, the very high visibility of these long-known results usually contributes to the fact that little attention is usually paid to them. So, one of the main features of nonlinear systems—the dispersion of the trajectories of the system depending on the scale of disturbances—almost has never been studied in application to the evolution of the physical and chemical systems. This evolution is ubiquitous in living systems at all levels of their organization. I call such processes with nonlinear dynamics occurring in a nonlinear medium "super nonlinear." It is these processes in viscoelastic media that cause hysteresis bifurcations of phase transitions, which leads to the "phenomenon of life." Although, by the way, in terms of the views set out in this book, life is not at all what is called a "phenomenon," i.e., unique; it is not a thermodynamic event.

In many studies scientists analyzed hundreds of possible chemical reactions in laboratories and their mathematical models in order to find a plausible method by which they could explain how molecules have evolved to produce the building blocks of life. But we must admit that so far we have not moved too much on the way to answering the seventy-year-old Schrödinger question: "What is life?"

In this book I did try to get a first iteration for solving this long-standing problem of a possible mechanism by which life may have originated in the primordial chemical soup that existed on early Earth or somewhere else in universe. My main idea is that by combining and recombining relatively simple molecular subunits such as nucleotides, amino acids, and lipids, it is possible to form ever-evolving viscoelastic matrices. As proposed in the book, a viscoelastic model of the origin of life indicates that these simple molecules could give rise to increasingly large and complex molecules, matrices, and networks of chemical reactions, presumably leading to life as we know it today. The major consequences of my analysis should be changes in our thinking about how life may have originated from pure chemicals. And in some degree I hope that my vision of the problem could help shed light on the problem and inspire new ideas in the field of developing new methods of experimental creation of some primitive life form in a laboratory. Identifying the systemic causes of life on the basis of the proposed ideas might help to significantly narrow the field for the experimental work carried out for the purpose of *in vitro* reproduction of life.

In many modern works, it is stated that, among others, the critical condition of life's emergence is the process of self-assembly of molecular structures. However, from these works it is not clear how evolution emerges and what is the physical selective factor of evolution. For example, scientists have created new kinds of particles, 1/100th the diameter of a human hair, that spontaneously assemble themselves into structures resembling molecules made from atoms [74]. These new particles "self-assemble" and form structures with different patterns. But this example shows that the "self-assembling" by itself does nothing. Proclaim, as many do, that "self-assembling" is the beginning of life, at least recklessly. By itself, this process does not determine the beginning and the direction of evolution.

Also, many researchers charmed by Prigogine's and others' work are trying to view life as a dissipative system. The main difference between living systems is that they can exist for long without the influx of external energy. However, in my opinion no one has demonstrated the existence of dissipative structures existing without a constant flow of energy. A dissipative structure is the simplest way to organize the flow of the mass and energy of inanimate matter, which is generally not associated with the phenomenon of life. Life as we know it is made up of multiple hysteresis loops at each hierarchical level. Inside these loops, metabolizing systems at a microlevel stay under the influence (at least partially) of the macrostate of the next system level. This makes it very difficult to untangle the simple cause-effect relationships that exist in the linear mechanics of the life sciences. It is mainly because the metabolism of all living systems is governed by renormalization in hysteresis loops serving as regulatory processes.

These regulatory processes determine the orientation of processes from one loop to the next. But the orientation does not exclude the branching processes in phase space of the metabolism in the form of the branching chains of chemical reactions with hysteresis loops.

These hysteresis loops include in themselves all feedback links. Renormalization of the ensemble probability of the particle states of the system in the hysteresis loops can be considered as an action of a feedback loops spectrum. The concept of feedback loops was developed by the founder of cybernetics, Norbert Wiener. The feedback loop in the Wiener sense is a system of conjugate elements in the circle, in which the initial impact has an effect on the next item, but sibling back in the ring also has an impact on the first. The result of such an organization is that the first link is affected by the next link in the chain, which means that the entire system of self-regulation, like the initial effect, is modified each time it passes the ring. Even in the simplest molecular living systems or gels, feedbacks have a spectral character, because the control of the process renormalization of the molecular ensemble consisting of a large number of different molecules with a large number of degrees of freedom based on a simple (single) feedback loop, generally is not possible. There are two main kinds of feedback loops with spectrum of intensity: negative loops to regulate inhibition; positive loops to amplify.

One can say that all living beings are dissipative structures, which have a tendency toward self-organization and pattern formation. But in the classical subject of nonequilibrium thermodynamics, in dissipative structures, flows of energy and matter and their pattern are determined by the physics of processes at this level of intensity of the processes and fields. And a higher level of structural hierarchy affecting it through a spectrum of feedback loops for inertial dissipative process does not exist. From the earliest days of the formation of biology, scientists have noticed that all life forms the mysterious ways of combining the stability of the structure with the flexibility of changes.

As well as dissipative systems, they depend on a constant flow of energy and matter that passes through them, but, in addition, and in contrast to the dissipative systems the living structures develop, reproduce, and evolve. The main difference between living systems from the dissipative structures in nonliving matter is that the flow of energy and matter in the structures of living systems are managed, at least in part, by the state of flow of energy and matter at the next higher level of the hierarchy. The criterion of hierarchy in this sense is a hysteretic nature of transitions between levels of living matter organization. Also, because all living matter responds to the influence of the environment through the changes in hysteresis loops, then these changes, in turn, affect its subsequent behavior. This means that living matter is structurally coupled, or in other words, is a learning system. While any system remains alive it will always "hysteresisly" adjoin another system. But in addition, it must also adjoin with the environment. This environment can be anything: the environment inside the cell structures in the cell, the extracellular environment of the cells, the surrounding organs of the body, the environment and the ecosystem of the body, apparently, and the social environment of social living systems.

Permanent structural change, adaptation in response to changes in external (boundary) conditions, defines the key characteristics of living systems. Due to the internal and external hysteresis conjugation, living systems are capable of evolution (and in particular to adaptation).

Furthermore, the viscoelasticity responsible for the emergence of life billions of years ago continues to manifest itself and have a significant impact on all the processes of mass transfer in living beings. The viscoelastic properties functionally manifest themselves in all parts of the body. All processes in the cells are determined not only by reactions in the viscoelastic matrix, but also the influence of viscoelasticity on the diffusion of molecules, the orientation of the interacting protein molecules, etc. All processes in the tissues and muscles are also determined by a viscoelastic medium. Our conception is determined by the viscoelastic properties of semen as a whole and all its parts, the egg and the uterine fluids involved, and the embryo.

Attempts to analyze the functioning of the organism at any level of the organization without viscoelasticity is doomed to fragmentation and the paucity of understanding of the real processes. But we must recognize that even for the most complete qualitative effects of viscoelasticity to the work of living systems we are only at the beginning of the beginning.

Proponents of intelligent design in the origin of life suggest that if natural selection is based on reproduction, there can be no Darwinian solution to the problem of life—the first living organism would be too complicated for Spontaneous Generation. Thus, if we think of the pre-life world as a chaotic accumulation of chemicals in the primordial ocean, "self-assembled" by chance in a primitive cell, the problem of life is a dead end. As a way out, some have offered vitalism, or God, and other variations on these themes. For example, some published studies assume a cognitive coherence in molecules. The idea is certainly very interesting, but compared with the parable of the Old Testament account of creation, it looks better for me personally. It's easier to believe in Santa Claus than in "thinking molecules."

But if you imagine that the key process of molecular evolution—the viscoelastic hysteresis hyper cycles of chemical reaction bifurcations, which include the reaction of replication—the mystery becomes a problem: how nonlinear astronomical, solar, lunar, seasonal, diurnal, geochemical, and molecular cyclic processes about four billion years ago led to the circumstances in which the viscoelastic hyper cycles were born.

Therefore, in science there is a good rule: when faced with unusual natural phenomena, seek answers to the known laws of nature—they are, in nine hundred and ninety nine cases out of a thousand the answers to the riddles of nature. And only if the answer is not found, you should look for it in the direction of the modification of the laws of nature. I hope that the proposed viscoelastic hypothesis of molecular evolution is included in those nine hundred and ninety nine answers, of which I spoke earlier. But more than that, further analysis shows that the key process of molecular evolution (viscoelastic renormalization of chaotic microstates by their final (future) macrostate) works well from the emergence of life to the present day life processes in living beings. At the moment, my dear reader, in the cells of your body, this viscoelastic renormalization process occurs in the trillions of trillions of times every second. And it means that when the limited number of external parameters achieve a certain range in the right chemical broth, life emerges, exists and evolves, and certainly is not a unique phenomenon. But the integrated universal characteristic of molecular biochemical evolution is the possibility of viscoelasticity emergence in this range of parameters. So I would suggest describing living things as a continuously evolving system of controlled chaos inside with viscoelastic feedback.

At the end of my work, I want to make another generalization, though a somewhat speculative remark. Viscoelasticity, perhaps only as a theoretical approximation of reality, or perhaps because of some deep laws of nature, manifests itself in an entirely different scale. So I mentioned earlier that the tectonically active planet can be seen as a viscoelastic body. A tectonic activity is considered by many to be one of the main conditions for the existence of early life on Earth.

But except for the scale of the planets in a strange way, viscoelasticity occurs in the deepest processes of matter and cosmological scales. So as shown in recent theoretical work [75] in sixes dimensional space, a black hole behaves like a viscoelastic material—fluid and a solid at the same time and it may generate an electrical field. Despite the fact that these effects are found only in the theoretical realm, the underlying equations could help scientists to find out and describe some of the real properties of the hot, extremely dense matter that existed immediately after the Big Bang. Theoretically, it can convert mechanical stresses into electric fields, and vice versa.

But if a black hole can be approximated by electro-viscoelastic materials then probably very similar laws have acted in the evolution of the early (quantum) universe and early (physicochemical) life forms. Moreover, from this point of view, the laws of evolution in the universe may have a degree of universality, which until now, scientists could not imagine. Of the theoretical constructs physicists know that as in the primitive chemical broth, in which the origin of life on Earth took place, in multidimensional universe a primordial four dimensional form of matter as well is consisted from fleeting and chaotic particles. That is, renormalization of quantum theory could determine the evolution of matter at this stage of its development.

Literature

1. Farid I. Zapparov, *The Food Delusion* (Charleston, SC, USA, 2012), ISBN-10: 1470061554.
2. James Lovelock, *The Revenge of Gaia* (Santa Barbara, CA, USA, 2006), ISBN 0-7139-9914-4.
3. 3. E. Schrödinger, *What is life?* (Macmillan, 1946).
4. G. M. Bartenev, C. Ya. Frenkel, *Polymer Physics* (Leningrad, Russia.: Chimiya, 1990).
5. G. Bistrai. ТЕРМОДИНАМИКА НЕРАВНОВЕСНЫХ ПРОЦЕССОВ В ОТКРЫТЫХ НЕЛИНЕЙНЫХ СИСТЕМАХ С ДЕТЕРМИНИРОВАННЫМ ХАОСОМ Диссертации на соискание ученой степени доктора физико-математических наук. Ekaterinburg, Russia, 2009.
6. A. Breitkreutz et al., "A Global Protein Kinase and Phosphatase Interaction Network in Yeast," *Science*, 2010; 328 (5981).
7. F. I. Zapparov. Влияние неньютоновских свойств жидкости на процессы конвективного теплообмена.- Диссертация...... канд. техн. наук.- Казань, Russia, 1982.
8. F. A. Garifullin, F. I. Zapparov. Конвективное движение в надкритической области для жидкости второго порядка. ИНЖЕНЕРНО-ФИЗИЧЕСКИЙ ЖУРНАЛ ТОМ 38, 6, 1980, Minsk, USSR.
9. G. E. Mikhailovsky. Отрицательная энтропия и диссипативные структуры, порожденные предельными циклами. Журнал физической химии. 1981. Т.55, №7. С.1877-1879, Moscow, USSR.
10. G. E. Mikhailovsky Биологическое время, его организация, иерархия и представление с помощью комплексных величин. http://www.chronos.msu.ru/RREPORTS/mikhailovsky1.pdf
11. L.P. Kadanoff, "Scaling laws for Ising models," *Physics* (Long Island City, N.Y., 1966) 2, 263.
12. K.G. Wilson, "The renormalization group: critical phenomena and the Kondo problem," *Rev. Mod. Phys.* 47, 1975, 4, 773.
13. Web Dictionary of Cybernetics.
14. I. A. Akchurin, "Basis of topological physics," in *Problems of Physics of Elementary Particles* (RAN, Moscow, 1994), 5-24.
15. R. Nigmatulin, "Fractional integral and its physical interpretation," *Theoretical and Mathematical Physics*, V. 90, № 3, 1992, 242-251, Moscow.
16. W. S. Childers et al., "Peptides Organized as Bilayer Membranes," *Angewandte Chemie International Edition*, 2010; DOI: 10.1002/anie.201000212
17. NASA Ames Research Center, "NASA Find Clues That Life Began In Deep Space," *ScienceDaily*, 31 Jan. 2001, web.

18. D. Walgraef, Spatio-Temporal Pattern Formation (Springer: Berlin, 1997).

19. S. Almagro et al., "Individual chromosomes as viscoelastic copolymers," *Europhys. Lett.*, 63 (6), 908–914 (Sept. 2003).

20. H. Tanaka, "Viscoelastic phase separation," *Journal of Physics*: Condensed Matter.2000;12:R207–R264.

21. V. Kumaran and G. H. Fredrickson," Early stage spinodal decomposition in visco-elastic fluids," *Journal of Chemical Physics*, 105 (18).8304-8313.

22. R. D. Gupta et al., "Directed evolution of hydrolases for prevention of G-type nerve agent intoxication," *Nature Chemical Biology*, 2011; DOI: 10.1038/nchembio.510.

23. H.C. Öttinger, "Dynamic Renormalization in the Framework of Nonequilibrium Thermodynamics," *Phys. Rev.* E 79, 021124.

24. A. Pross, "The Driving Force for Life's Emergence: Kinetic and Thermodynamic Considerations," *J Theor. Biol.*, February, 2003, 7;220(3): 393-406.

25. R. M. Turk, N. V. Chumachenko, and M. Yarus, "Multiple translational products from a five-nucleotide ribozyme," *Proceedings of the National Academy of Sciences*, Published online February 22, 2010 DOI: 10.1073/pnas.0912895107

26. Y. Gambin et al., "Visualizing a one-way protein encounter complex by ultrafast single-molecule mixing," *Nature Methods*, 2011; DOI: 10.1038/nmeth.1568

27. J. D. Sarfati, http://www.origins.org.ua/page.php?id_story=198

28. S.L. Miller and A. Lazcano, "The origin of life-did it occur at high temperatures?" *J. Mol. Evol.*, 41:689-692, 1995.

29. Yu. L. Klimontovich,, Введение в физику открытых систем (Moscow, Russia, 2002), Янус-К.

30. W. Lee et al., "Dynamic self-assembly and control of microfluidic particle crystals," *PNAS*, December 13, 2010 DOI: 10.1073/pnas.1010297107

31. C. J. Delebecque et al., "Organization intracellular reactions with designed RNA," *Science*, July 22, 2011;333(6041): 470-4.

32. J.A. Pojman, "Self Organization in Synthetic Polymeric Systems," *Annals of the New York Academy of Sciences*, V. 879, 1999: 194–214. doi: 10.1111/j.1749-6632.1999. tb10420.x.

33. A. Loskutov and A. Mikhailov, http://chaos.phys.msu.ru/loskutov/PDF/Loskutov. pdf

34. Michigan State University, "Studying the evolution of life's building blocks," *ScienceDaily*, February 20, 2012, web.

35. L. Burroughs et al., "Asymmetric organocatalytic formation of protected and unprotected tetroses under potentially prebiotic conditions," *Organic & Biomolecular Chemistry*, 2012; DOI:10.1039/C1OB06798B.

36. University of California–Santa Cruz, "Synthetic Biology Yields Clues to Evolution and Origin of Life," *ScienceDaily*, February 15, 2009, web.

37. J. Sarfati, "Origin of life: the polymerization problem," http://creation.com/origin-of-life-the-polymerization-problem

38. R. Vegners et al., "Use of a gel-forming dipeptide derivative as a carrier for antigen presentation," *Journal of Peptide Science*, Volume 1, Issue 6, November/December 1995, 371-378.

39. V.I. Goldanskyi and V.V. Kuzmin, "Spontaneous breaking of mirror symmetry in nature and the origin of life," *Sov. Phys. Usp.* 32 1–29 (1989); DOI: 10.1070/PU1989v032n01ABEH002674

40. Pierre de Marcellus et al., "Non-racemic amino acid production by ultraviolet irradiation of achiral interstellar ice analogs with circularly polarized light," *The Astrophysical Journal*, 2011; 727 (2): L27 DOI: 10.1088/2041-8205/727/2/L27

41. K. Zhao et al., "Local chiral symmetry breaking in triatic liquid crystals," *Nature Communications*, 2012, 3: 801 DOI: 10.1038/ncomms1803

42. University of Gothenburg, "Chirality: New method to consistently make left-handed or right-handed molecules," *ScienceDaily*, June 18, 2011, web.

43. California Institute of Technology, "First artificial neural network created out of DNA: Molecular soup exhibits brainlike behavior," *ScienceDaily*, July 20, 2011, web.

44. S. P. Mahal et al., "Transfer of a prion strain to different hosts leads to emergence of strain variants," PNAS, 2010; DOI: 10.1073/pnas.1013014108.

45. D. D. Leipe et al., "Did DNA replication evolve twice independently?" *Nucl. Acids Res.*, 1999, 27 (17): 3389-3401. doi: 10.1093/nar/27.17.3389.

46. C. Violle D. R Nemergut, Z. Pu,L. Jiang, "Phylogenetic limiting similarity and competitive exclusion," *Ecology Letters*, 2011; DOI: 10.1111/j.1461-0248.2011.01644.x

47. R. E. Beardmore et al., "Metabolic trade-offs and the maintenance of the fittest and the flattest," *Nature*, 2011; DOI: 10.1038/nature09905.

48. University of Florida, "Driving Force Of Evolution? Evolution Of Proteins Linked To Species' Metabolic Rate," *ScienceDaily*, October 8, 2007, web.

49. Public Library of Science, "Natural Selection May Not Produce The Best Organisms," *ScienceDaily*, July 21, 2008, web.

50. D. R. Schrider et al., "Pervasive Multinucleotide Mutational Events in Eukaryotes," *Current Biology*, June 2, 2011 DOI:10.1016/j.cub.2011.05.013.

51. P. P. Bhat et al., "Formation of beads-on-a-string in break-up of viscoelastic filaments," *Nature Physics*, 2010.

52. Barnosky et al., "Has the Earth's sixth mass extinction already arrived?" *Nature*, 471, 51–57, March 3, 2011, doi:10.1038/nature09678.

53. P. L. Indovina et al., "Thermal hysteresis and reversibility of gel–sol transition in agarose–water systems," *J. Chem. Phys.* 70, 2841, 1979; http://dx.doi.org/10.1063/1.437817

54. R. Youshida et al., "Design of novel biomimetic polymer gels with self-oscillating function," *Science and Technology of Advanced Materials*, 3 (2002), 95-102.

55. O. Kuksenok et al., "Exploiting gradients in cross-link density to control the bending and self-propelled motion of active gels," *J. Mater. Chem.*, 2011, 21, 8360-8371. DOI: 10.1039/C0JM03426F

56. http://www.youtube.com/watch?v=_EQY2x7avWo&feature=related

57. S. Maeda et al., "Active Polymer Gel Actuators," *Int. J. Mol. Sci.* 2010, 11, 52-66; doi:10.3390/ijms11010052

58. S. Maeda et al., "Design of Autonomous Gel Actuators. Polymers, 2011, 3, 299-313; doi:10.3390/polym3010299.

59. A. Jha, "Pulsations reveal which embryos have the best chance of success in IVF," http://www.guardian.co.uk/science/2011/aug/09/pulsations-embryos-success-ivf

60. A. Ajduk et al., "Rhythmic actomyosin-driven contractions induced by sperm entry predict mammalian embryo viability," *Nature Communications* 2, article number: 417. doi:10.1038/ncomms1424.

61. R. S. McIsaac et al., "Does the Potential for Chaos Constrain the Embryonic Cell-Cycle Oscillator?" *PLoS Computational Biology*, 2011; 7 (7): e1002109 DOI:10.1371/journal.pcbi.1002109.

62. T. Kiyomitsu and I. M. Cheeseman, "Chromosome- and spindle-pole-derived signals generate an intrinsic code for spindle position and orientation," *Nature Cell Biology*, 2012; DOI: 10.1038/ncb2440.

63. Rockefeller University Press, "How The Cell Finds Its Center," *ScienceDaily*, April 17, 2001, web.

64. J. Ranft et al., "Fluidization of tissues by cell division and apoptosis," PNAS, December 7, 2010, vol. 107 no. 49 20863-20868.

65. S. Kohler, V. Schaller, and A. R. Bausch, "Collective dynamics of active cytoskeletal Networks," Technische Universitat Munchen, Germany, May 2011, http://arxiv.org/PS_cache/arxiv/pdf/1105/1105.4475v1.pdf

66. Vanderbilt University Medical Center, "New Molecular Mechanism Associated With Arrhythmias Discovered: Possible Novel Target For Treating Arrhythmias" *ScienceDaily*, January 22, 2009, web.

67. S. Busch et al ., "Molecular Mechanism of Long-Range Diffusion in Phospholipid Membranes Studied by Quasielastic Neutron Scattering," *J. Am. Chem. Soc.*, 2010, 132 (10), 3232–3233, DOI: 10.1021/ja907581s.

68. American Institute Of Physics, "Spiral Waves Break Hearts: Importance Of Communication Between Cardiac Cells Is Demonstrated," *ScienceDaily*, February 11, 2002, web.

69. H. Karcher et al., "A Three-Dimensional Viscoelastic Model for Cell Deformation with Experimental Verification," *Biophysical Journal*, Vol. 85, November 2003, 3336–3349.

70. Georgetown University Medical Center, "For The First Time, Patterns Of Excitation Waves Found In Brain's Visual Processing Center," *ScienceDaily*, August 4, 2007, web.

71. G. Guigas et al., "Probing the Nanoscale Viscoelasticity of Intracellular Fluids in Living Cells" *Biophysical Journal*, Vol. 93, Issue 1, 316-323, July 1, 2007.

72. A. Blumenfeld, Биофизика N 1. C. 129, (Moscow, Russia, 1993).

73. Y. Jamali, M. Azimi, and M.R.K. Mofrad, "A Sub-Cellular Viscoelastic Model for Cell Population Mechanics," PLoS ONE 5(8), 2010: e12097. doi:10.1371/journal.pone.0012097.

74. Yu. Wang et al., "Colloids with valence and specific directional bonding," *Nature*, 2012 DOI: 10.1038/nature11564.

75. J. Armas et al., "Black Branes as Piezoelectrics," *Phys. Rev.* Lett. 109, 241101 (2012).

76. R. Youshida et al., "Design of novel biomimetic polymer gels with self-oscillating function," *Science and Technology of Advanced Materials*, 3 (2002), 95-102.

77. O. Kuksenok et al., "Exploiting gradients in cross-link density to control the bending and self-propelled motion of active gels," *J. Mater. Chem.*, 2011, 21, 8360-8371. DOI: 10.1039/C0JM03426F

78. http://www.youtube.com/watch?v=_EQY2x7avWo&feature=related.

79. A. T. Winfree, *The Geometry of Biological Time*, (Springer, Berlin, 2001).

80. A. Fritsch et al., "Are biomechanical changes necessary for tumour progression?" *Nature Physics*, 6, 730–732 (2010) doi:10.1038/nphys1800.

81. J. Szymański et al., "Diffusion and Viscosity in a Crowded Environment: From Nano- to Macroscale," *Journal of Physical Chemistry B*, 2006; 110 (51).

82. http://www.dfhcc.harvard.edu/

83. G.J. Jing et al., "Aberrant expression of nuclear matrix proteins during HMBA-induced differentiation of gastric cancer cells," *World Journal of Gastroenterology*, 2010; 16 (17).

84. P. Glansdorff, I. Prigogine, *Thermodynamics Theory of Structure, Stability and Fluctuations* (London: Wiley-Interscience).